A New Framework for IT Investment Decisions

A practical guide to assessing the true
value of IT projects in business

by Antony Barnes

HARRIMAN HOUSE LTD

3A Penns Road
Petersfield
Hampshire
GU32 2EW
GREAT BRITAIN

Tel: +44 (0)1730 233870
Fax: +44 (0)1730 233880
Email: enquiries@harriman-house.com
Website: www.harriman-house.com

First published in Great Britain in 2010 by Harriman House.

978-0-857190-26-0

British Library Cataloguing in Publication Data
A CIP catalogue record for this book can be obtained from the British Library.

Printed in the UK by CPI Antony Rowe, Chippenham.

Contents

About the Author

Antony Barnes is an IT consultant with over 15 years of experience. With a background in system design and programming, he has a strong track record of success in helping companies gain a competitive advantage through understanding and maximising the value of their IT systems. Antony currently works in the area of mergers and acquisitions, helping businesses to maximise the value of their IT investment throughout the deal process. He holds a BSc and MSc from Imperial College, London, and an MBA from Warwick Business School.

List of Abbreviations

API: Application programming interface – the standard definitions and software components related to a system that enable programmers to write software which connects with that system.

BAU: Business as usual – a commonly-used term which refers to the normal operations of a company (as opposed to programme activities designed to change or improve the operations).

CAD: Computer-aided design – software which helps designers to create and manage their designs on a computer rather than on a drawing board.

Capex: Capital expense – the costs associated with investing in business improvements.

CEO: Chief Executive Officer – the most senior executive manager within a business.

CFO: Chief Financial Officer – the head of finance within a company.

CIO: Chief Information Manager – the head of IT within a business.

COO: Chief Operating Officer – the head of operations within a company.

CRM: Customer relationship management – software which allows a company to track and maximise the value of its customer relationships.

EDI: Electronic data interchange – technologies and standards which allow companies to communicate and conduct business between themselves electronically, for example by sending purchase orders automatically.

ERP: Enterprise resource planning – software designed to work as a single system across an enterprise, breaking down the barriers between functional areas.

IT: Information technology, taken here to include both technological and systems (i.e. to include what is sometimes referred to as information systems, or 'IS').

M&A: Mergers and acquisitions – also known as takeovers, where one business or group of investors takes over or merges with another.

NPV: Net present value – a calculation designed to establish the present value of a proposed investment, taking into account the fact that future cash flows should be assigned less value than current ones by means of a discount factor.

Opex: Operating expense – the costs of keeping a business running.

ROI: Return on investment – has no well-defined and generally accepted meaning. In this book it is taken to mean a clearly identified financial return on a capital investment.

SAP: *Systeme, Anwendungen und Produkte in der Datenverarbeitung* – a well-known ERP system, and also the company that designs and sells it (SAP AG, based in Germany).

SG&A: Sales, general and administrative – the overheads for running a business.

TCP/IP: Transmission control protocol/internet protocol – a widely-used communications protocol, which is also the standard on which the internet is built.

All trademarks in the text are acknowledged.

Introduction

This book sets out a new conceptual framework for making decisions about investments in information technology (IT). It aims to help business and computer people make such decisions in a consistent way, in order to re-assess a company's IT project portfolio.

The new framework offers some new insights on information systems, leading to conclusions that are very different to the currently accepted wisdom. For example, it suggests that not every IT project needs to demonstrate a return on investment. Computer systems usually play a supporting role in business, and it makes little sense to try and justify the costs of IT projects in isolation. Nor should IT strategy necessarily be aligned with business strategy – the framework shows that the planning cycles for a business and its computer systems are not always closely synchronised. Effectively, the framework offers a new perspective on what an IT strategy really is. The central strategic question that IT people face is about software applications – what applications will be provided to a business, and when.

Let me clarify what this book is not: it is not about the details of system selection, or security, or IT governance, or the many other challenges that are faced by an IT team. These are better understood by others. Nor have I covered the implementation of projects. It is well known that IT projects have a high rate of failure – one can pick up a newspaper almost any day and find a story about a failed computer system. (Some of the most high-profile cases are found in the public sector, and many people draw the conclusion that managers in the public sector are no good at information technology. I disagree: I think there are just as many failed IT projects in the private sector, but businesses are better able to keep their mistakes secret.) The focus of this book is on the IT investment decisions which businesses need to make, so it does not cover the project life-cycle from the investment decision onwards.

In particular, this book is not about the costs of information systems. It is fair to think of any business decision, including a decision about such systems, as being a balance between costs and benefits. But I think the costs of information systems are reasonably well understood, even if they are often difficult to calculate. I have little to add to this debate, so rather than repeat what others have said, I have not included any detailed analysis of IT costs. The benefits, on the other hand, are what this book is all about.

Although aimed primarily at businesses, I hope that many of the ideas put forward here will be equally useful within a public-sector or not-for-profit organisation. Technology investment is complex, whether in the private or public sectors, and although the competitive pressure which businesses face is not usually felt by public-sector or not-for-profit organisations, that pressure is often replaced by a constant drive to reduce spending.

This book is organised as follows. Part I sets out the problem, which is the lack of a consistent framework for assessing IT value. If you already agree that there is such a problem, you might prefer to skip this argument and go straight to Part II, which is a presentation of a new decision framework, summarised at the end of Chapter 5. In Part III, the framework is applied to a range of real-life projects, and from this are derived four types of benefit that IT projects can offer. The details are summarised in a table at the end of Part III. Part IV brings the discussion to a close with some suggestions on how to measure the value of the four types of benefit, and how to prioritise an IT budget across different projects.

Examples and case studies are included where they are helpful in clarifying points, and because they indicate how companies make decisions about IT spending. I have also relied heavily on comments made by companies' chief information officers (CIOs), in their annual reports, in presentations to investors or in interviews with the IT press. In general, I think it fair to assume that these comments can be taken as representing a company's thinking when it commits to an IT project. But I will not pretend that these examples, case studies and CIOs' comments provide hard evidence for the points I have made. This is not an academic treatise, but a book that presents a model for managers – and hopefully a useful one.

There is one final reason for including statements from companies' CIOs. Over the years, IT people have learned many lessons the hard way, through practical experience and plentiful mistakes. In doing so, they have developed a collective wisdom about their field, and that is why I have included their views. My personal belief is that CIOs are making better investment decisions today than in the past, and that their practical decisions are ahead of the theory. This book is an attempt to bridge that gap.

Part I
The Problem With ROI

This section sets out the problem that businesses face with ROI (return on investment), namely the lack of a consistent framework for making decisions on IT investments. Later sections will present a new decision framework, and show how it can be applied to IT projects.

At the moment, businesses are advised to use cash measurements of value when making IT investments, just as for any other capital project. But these measurements do not work well for IT projects, a situation that needs to be addressed. Worse, a focus on cash returns may lead managers to spend in the wrong areas.

Part I also presents some possible reasons why IT projects are hard to value. IT systems are highly complex, but they are also (relatively) cheap. Therein lies much of the problem.

Chapter 1
Hard Benefits Are Hard to Find

How do you decide whether or not to spend money on computer systems? If you are one of those people who has to make or recommend such decisions, you will probably have a ready answer: you might talk about return on investment, or business cases, or hard and soft benefits. But I suspect that most business people will not come up with a clear response.

Any investment decision is a balance between costs and benefits. The problem comes when you can measure one side of the balance (the costs) but not the other (the benefits). It would be convenient if it were possible to work out the benefits of investments in IT, but they are hard enough to identify, let alone to quantify. Companies must grapple with unknown benefits and unpredictable outcomes, because computer systems are inherently complex. Contrary to popular belief, all the benefits of an information system cannot usually be predicted ahead of time, nor can the financial value of those benefits be calculated.

The danger is that companies will focus on the costs (because they are measurable), and ignore the benefits (because they are not easily measurable). Perhaps part of the reason for this is that there is no standard framework for analysing benefits. Contrast this with accounting: all accountants know and understand the basic building blocks of their trade, such as debits and credits, profit and loss statements, balance sheets, cash flows, and so on. But there is no corresponding framework for the benefits of computer systems, and that makes it hard to decide on IT investments.

Lacking a framework for assessing the benefits of computer systems, firms have instead relied on a 'project-gating' approach. This process is based on a series of decision 'gates', and a project is only approved if it passes through all the gates. Requests for new projects can come from users, from management or from the IT team. The gating process normally hinges on a business-case document, which includes details such as a project summary, a list of the advantages, an analysis of the risks, a rough timescale, an estimate of resources and an analysis of dependencies on other projects. The benefits are set out in terms of 'soft' and 'hard' benefits. The soft benefits are such intangibles as better information or improved staff morale. The hard benefits are measured in cash, at least in theory.

While at first glance this may seem a reasonable approach, it fails to allow the firm to take an overall view of its IT project portfolio. There is no underlying model that allows it to say whether or not a project 'fits' into the portfolio, but only an ad hoc set of rules which are applied at each gate. So the project-gating approach encourages managers to view each project in isolation. For example, a new automation system might be evaluated one day; a network upgrade the next; a server-virtualisation project after that, and so on. The problem with the project-gating approach is that it is fundamentally concerned only with the *process* of selecting projects. It has nothing to say about the much more important matter of maximising the overall value of the IT investment to the business. By contrast, the framework to be presented in this book suggests that decisions on technology projects should be made by considering the project portfolio as a whole.

The Unquestioned Logic of ROI

At the core of the project-gating approach is the concept of return on investment (ROI). For the purposes of this book, ROI is taken to mean some sort of financial appraisal based on projected cash flows, calculated from the bottom up. In practice this is likely to be a combination of net present value (NPV), internal rate of return and payback time. ROI and its variants are the evaluation techniques to be found in the accounting textbooks,[1] and in theory should work across all projects, not just those involving computers, because all projects should generate positive future cash flows. ROI offers a single yardstick to measure the value of any project.

It has become a kind of mantra to say that 'an ROI must be demonstrated' for all IT projects. It is so much an accepted norm that people no longer even question the logic. Paul Strassmann set out the case for a much more rigorous analysis of IT costs and benefits in his book *The Squandered Computer*, where he said: "Every computer project proposal should demonstrate the discounted cash flow of its proposed business improvement."[2]

The problem in the context of IT projects lies in working out the numbers. There is a hidden assumption in the ROI calculation, which is that *it is possible to calculate future cash flows*. In the case of most IT projects this assumption is simply not true. Although it is usually possible to work out the costs of IT projects, at least to a reasonable degree of accuracy, it is much more difficult to calculate hard numbers for the future benefits. This is because the fundamental role of information technology is to *support* the operational core of the business. In the most general terms, IT systems do not modify the cash flows through a business, but merely support activities which do.

Take the simple example of a taxi business that is deciding whether to invest in a new taxi. It ought to be straightforward to work out the ROI, because the costs and the potential benefits are reasonably predictable. The only unknowns are the commercial considerations – whether there will be enough potential customers to justify the cost of a new taxi.

But suppose the company is considering an investment in a computerised accounting system. Is it still possible to work out an ROI? Where are the hard cash savings or advantages? I have no idea. And I don't believe anybody else has any idea either. A new accounting system will undoubtedly offer benefits, but these will not be easy to measure.

Of course, there will be *some* hard benefits that come from investing in the accounting software. For example, the managers might be better able to control the company's working capital, because they have a clearer view of stock, accounts payable and so on, which would translate into a benefit. On that basis, and adding in some other hard benefits, it might be possible to work out an ROI. But the hard benefits do not represent *all* the benefits of the system, in the way that the ROI for a new taxi does. And if the ROI calculation only identifies a fraction of the full benefits, how can it be used to compare IT projects with others where ROI *does* identify the full benefits (such as a new taxi)?

Infeasibility of calculating the full benefits

There is another problem with calculating the benefits of projects using ROI (at least with a 'bottom-up' calculation): while the benefits must be calculated at the detailed process level, the decision must be made at the top level of the business. One of the biggest IT decisions that a business can make is to implement an enterprise resource planning (ERP) system. The cost of such systems can run to tens or hundreds of millions of pounds, and a decision of that magnitude must be made at board level. But the eventual benefits will be gained at the front line, by changing one process at a time, day by day, inch by inch. In order to calculate the full benefits, the firm would need to go through *all* the affected processes, and work out how they will be improved. This is not feasible for normal decision-making timescales. It might be argued that such a limitation is true of many other types of project; but it applies particularly to IT projects because of their very nature as agents of change, in that most of the benefits can only be derived by doing things differently. A new taxi does not change the way a taxi business operates, but a new accounting system does, and it is hard to calculate the benefits at a detailed level.

This also makes it hard to compare investments in IT systems with other capital projects where the benefits *can* be calculated. For example, a film studio might be considering the production of a new blockbuster movie at the same time as it considers a major investment in a new IT system. A calculation of the ROI for the film is reasonably straightforward: it involves estimates of the costs and of the ticket sales, plus any ancillary revenues from product endorsements, and so on. The calculation is complex but feasible, and the unknowns are primarily concerned with the likely popularity of the movie. It is also relatively simple to measure the success of the movie after it is released, since there is a clearly identifiable revenue stream (actually, it is usually possible to predict the success or otherwise of a movie after the weekend it is released). But none of this relative simplicity can be found in the case of an IT investment, so to compare the financial viability of the two projects is a matter of chalk and cheese.

And there is another fundamental reason why ROI is a poor yardstick for assessing IT spending: many IT projects are not even intended to generate an increase in cash flows. Disaster recovery programmes are a good example: when a business invests in a back-up data centre, it does not expect to see a financial return; rather, it wants to protect its existing business and revenues. ROI cannot be used to assess an investment in the provisions required for disaster recovery, because there will be no return on the investment.

The reality is that *ROI is not a useful measure for the vast majority of IT investments*. The mantra that 'an ROI must always be demonstrated' is, in the IT context, quite simply wrong. At the time of writing, it nonetheless remains the consensus view.

Stretching a definition

Given the problems with ROI, some IT people stretch its definition to include intangible benefits. A typical comment might be: 'The ROI on this project will be an improvement in staff morale and customer service.' Sure enough, when a business says 'we must always show an ROI for our IT projects', and then redefines ROI to include intangible benefits, it can indeed demonstrate an ROI. Taken to an extreme, calculating an ROI for a project then becomes synonymous with building a business case. But that is very different from what accountants and financial people mean by ROI. They mean something quite specific – the expected return on an investment, measured in financial terms.

Any other definition is apt to cause confusion. It seems better to talk about ROI in the same way as an accountant would understand it, and to refer to other, intangible, benefits as being part of a business case.

There is of course no suggestion here that a business can simply forget about ROI – it must give a decent return to its shareholders and debt holders if it wants to stay in business. To do that, it must use its funds to invest in projects that will generate a return higher than the cost of its capital. But the key is to understand the likely timescale for that return. In the case of an information system, this might be a very long time. In other words, an information system *might* actually generate a cash return over the long term, but the benefits are so hard to assess, and so mixed in with other factors, that ROI is not a useful measure in the short or even medium term.

When ROI drives efficiencies – but not efficient business

In fact, a rigid insistence on demonstrating ROI can focus the IT budget on the wrong areas. This is because a positive ROI *can* be calculated for some technology projects. Unfortunately, this is almost too easy. It is very commonly the case that when an investment is made in a technology platform, the ongoing cost of maintaining that platform is reduced. In other words, by investing some money now, the business will save money over the longer term, and the total costs to the business taken over a time span of five years, say, are lower. Such projects lead to a reduction in the overall direct cost of IT.

For example, many businesses (at the time of writing) are considering the use of 'virtualisation' software. This technology allows for many virtual servers to be run on one machine. As well as gaining advantages in flexibility and resilience, it is possible to reduce the overall costs of maintaining the servers, simply by replacing many machines with one. From an accounting perspective, there is a positive ROI. But has the business gained real value? Not really. All that has happened is that the cost of providing existing IT services has been reduced. Given a limited pot of investment funding, it does not necessarily make sense to focus on projects which have a positive ROI, but which soak up the IT budget.

Arguably, technology teams spend too much of their budgets on projects that demonstrate clear savings and therefore a positive ROI, but which fail to move the business forward. Such projects are good internal politics, especially if a firm is pressing for savings across the board, but not necessarily good business.

Spending the IT budget on cost-saving initiatives means there is less money left for genuinely beneficial projects which might sharpen the firm's competitive edge. Of course, if a company views IT simply as a support function, then this simple type of IT cost-cutting project is the best that can be done; and in that case, the ROI really *is* the total benefit, because the business is not changing at all. The direct IT cost savings are identical to the overall benefit of the project, because there are no ancillary soft benefits.

ROI: Broken in Theory and in Practice

In summary, ROI is not a useful tool across the full range of IT projects, given the impossibility of calculating all the related cash flows. But if it is not useful for all projects, how useful is it for any? The whole point of the ROI approach is supposed to be that different projects can be compared, because there is a method of measurement common to them all. But how does one decide between two projects if they cannot be measured by the same yardstick?

ROI at least has the advantage of being conceptually simple. But we cannot solve complex problems by pretending that a simple tool will do the job. We need to grapple directly with that complexity and come to an informed decision.

Actually, businesses don't follow the mantra about ROI being the only measure (if they did, they would never invest in any IT systems). Business people understand that ROI won't give them all the answers they need. In fact, any sensible business manager will usually be happy to fall back on soft benefits as a justification for a project, and to forget about ROI. Rather than relying only on ROI, business people use their instinct and sense of where the business needs to move.

In an interview with *Computerworld* magazine, Intel's CIO, John Johnson, said:

> *It's not always easy to predict how you would even do an ROI analysis. You could spend a year figuring out ROI, and then you might have wasted a year. [But] you do need to figure out what the business value proposition is.*[3]

Intel is a sophisticated business that really understands technology – and it does not insist on demonstrating ROI for its IT projects. In fact, a 2006 survey of businesses in the UK found that in such cases 89 per cent of them did not measure ROI[4] – and it would be surprising if the 11 per cent who did were actually measuring anything useful.

What usually happens with an IT project is something like the following: a group of project sponsors come up with an idea for a new IT innovation. They know the project makes sense, but are unable to provide a clear ROI for the required investment. Nonetheless, they put forward a solid business case for the project, which typically includes a mixture of tangible and intangible benefits, and the business decides to approve it anyway. Of course, the danger with the intangible benefits is that they are unstructured and can be whatever the management team wants; they might include such effects as higher customer satisfaction, or better internal communications, or even improved staff morale. In fact, every company looks at a different set of intangible benefits, because there is no underpinning framework – and this is a major part of the problem.

Nevertheless, I believe these decisions are right as often as they are wrong, and that the practical decisions are well ahead of the theory. This book is an attempt to redress the balance, and to provide a theoretical framework that can support better decision-making.

Bridging a gulf of misunderstanding

At a minimum, management teams should be able to hold an informed debate about their IT investments. At the moment, there is often a gulf of misunderstanding separating business and technology people: the former see revenues and costs, while IT people see systems and technologies. On their own terms, both sides are right, but it is as if they are talking two different languages. They have no common framework that would allow them to conduct a sensible debate about the value of information systems.

Chapter 2
The Complex World of Computers

If ROI does not work, how else can decisions be made about computer systems?

We have grappled with the problem of estimating benefits for decades, since the earliest computers were invented. In those days, computers were mainly used to automate the most repetitive tasks. This sort of data-processing took away some of the drudgery of clerical work – people sat in front of simple terminals connected to powerful mainframe computers, and carried out routine tasks. Because computer systems at that time were so (relatively) simple, it was correspondingly simple to measure their value. When a manual process was automated via a computer, it was straightforward to calculate the hours worked, the time taken to process per unit, and so on. This allowed companies to work out how much the system was saving in labour costs, and thus to calculate its value.

The PC revolution

But during the 1980s, the personal computer (PC) appeared, and people started to use computers in a very different way. Instead of carrying out mundane clerical tasks, they could do their own work, at their own pace, with their own programs, and make their own decisions. Looking back, it is hard to comprehend fully the effect of the PC revolution on people's working lives, and the shift in power that it brought about.

One of the most exciting applications of the time was the spreadsheet, which allowed people to perform complex calculations on their own PCs. We are accustomed to thinking about the effects of computers in terms of speed, but the effect of the PC revolution was also to put computing *power* in the hands of people. Equipped with a PC and a spreadsheet, people were able to work independently, and they had no need for the help of their IT department. In his book *The Alignment Effect*, Faisal Hoque described the effects of this change on the financial services industry's analysis of investment models:

> *But the real revolution that the spreadsheet kicked off wasn't just about efficiency and automation. By unburdening analysts from the pedantic work of manual calculations, spreadsheets lowered the marginal cost of evaluating new scenarios from thousands of dollars to almost zero. This, in turn, encouraged experimentation and creativity. The same employee who once spent days perfecting a single model could suddenly produce several alternatives in a single afternoon.*[5]

This changed the calculation about the value of computer systems. As long as people were doing routine tasks on dumb terminals, it wasn't hard to work out how much time was being saved. But when they started working in new ways on their own PCs, those calculations became much harder.

If the PC revolution put power into the hands of users, the next step in the story of computers was the advent, in the late 1980s and early 1990s, of client-server computing on a large scale. Of course, people had already been using central fileservers to store files from their PCs. But this was (and is) a rather basic task for a server – essentially, using a fileserver as simply an electronic filing cabinet. With client-server systems, however, a desktop PC works together with a server: the PC displays information and allows the user to perform local processing, while the server manages data and queries. Client-server systems further complicated the calculations of value, and the split between central and personal processing thus made the calculation of ROI even harder.

In the mid-1990s the internet emerged as a global phenomenon. Though it and its predecessors had been born as long ago as the 1970s, it was the invention of the worldwide web, by the scientist Tim Berners-Lee, that really brought the internet to the world. Together with email, the other key internet application, the web turned the internet into an international phenomenon. And once again, the valuation of information systems became yet more complex, because PCs were now connected to an information repository and communications network that spanned the globe. This had enormous value for modern businesses, but to assign a hard ROI number to that value is virtually impossible.

Improved technology management and construction

As computer systems have become global and more complex, the IT world has worked hard to manage this complexity. At the technical level, we have much more powerful tools than ever before. New object-oriented languages such as Java and C# have made programming a much more robust and reliable process, while management tools have made it possible to control a global corporate network from a single console, and to manage a PC from the other side of the world.

In parallel with these improvements in technology, technologists have also worked hard to build better processes for building and managing computer systems. In particular, much effort has been expended in improving the way that software is developed – because of the problems that people have encountered in building software that actually does what a business wants it to do. The traditional way of developing software has come to be called the 'waterfall' approach. This attempts to document the user requirements for a software system in such detail that any ambiguity is eliminated. Even though IT people and business people often appear to talk two different languages, they should at least be able to agree on a written contract that they both understand. To that end, user requirements are documented in detail and signed off by the users. These requirements are then converted into a software design, which in turn is used by programmers who convert the design into programs. Finally the software is tested, approved and launched.

Although this is an outwardly logical approach, it can be cumbersome, and also tends to generate a huge amount of documentation. In addition, it depends on business people having a full understanding of everything they want, right from the start of the project. In reality this is rarely the case – because the people concerned do not have full knowledge of what they want, or because the opportunities to automate processes only become clear when people see the software itself, or simply because the business has moved on by the time the software is ready.

Recently, the waterfall approach has tended to be superseded by more dynamic processes. These 'agile' approaches are based on an iterative model, whereby the business users are not expected to understand and document all their requirements right at the start. Instead, a prototype is built quickly by the technology team, and the users are able to work with it and change it. Then another iteration is produced, and so on, until the software does what the business wants. The approach is based on the programmers working interactively with the users and building a level of trust with them.

Unimproved technology decisions and missing value estimations

Using such approaches, improvements have been made in building software quickly, while at the same time meeting the business people's requirements. Nevertheless, it only allows technology people to work better at the level of the technology *system* – it does nothing to help the business make intelligent decisions about what systems to develop in the first place. In other words, all the improvements in software engineering will help to build better software, but not to decide what software to build. Also, these software-engineering approaches only apply (obviously) to software-engineering projects. They cannot be used to make decisions on technology-infrastructure projects, for example. So for a busy CIO, who has to prioritise many different types of project, these software-engineering methodologies are only part of the answer.

There is a deeper problem, however, with both the waterfall and agile approaches, and, once again, it has to do with the estimation of value. Even if the users are successful in analysing all the functions that will be required of a software application, they have still said nothing about the *value* of these functions. To take the example we looked at earlier, that of an accounting system for a firm of taxis: one of the functions the users might want is a report of monthly takings per driver, and so they would specify the details of such a report in their user-requirements document. But what they will have done is say *what* they want, not the *value* of it to the business. To do this, they would have to say how they currently measure takings per driver; how that will change with the new report; and how much they estimate the change will be worth. But as Paul Strassmann says in *The Squandered Computer*, they really need to compare the savings with the best possible *manual* process. He recommends that managers "calculate the returns on computer investments as the difference between the costs of automation and the next least expensive solution".[6] He also says:

> The correct ROI is the cost advantage of performing the identical functions after making improvements by other means. The benefit of renting a parking space for an automobile in New York City is not the avoidance of police parking fines: it is the difference in the cost from the next least expensive parking option.[7]

It is hard enough to specify requirements, without then trying to estimate their value. In reality, most businesses do not even attempt to calculate these numbers.

Meanwhile, improvements have also been made in project management, by attempting to capture the key elements of what good project managers do, and then recording them in a template for project management – an example is the PRINCE2 standard.[8] While there is no doubt that these are useful tools for an IT team, whatever their success they can only affect a firm's ability to define and execute projects; they cannot help with the core decision about what projects should be considered in the first place.

The false hope of better communications

We have considered improvements in computer technology and processes, and now need to complete the picture by talking about computer people: have they got any better at establishing and communicating the value of computer systems?

Over the years, IT professionals have engaged in a somewhat self-critical debate, the general theme of which has been that they are too involved with technology and not enough with business. This has led to the idea that computer people need to become business people, talking the latter's language. To an extent this is a helpful suggestion, because it draws the IT team into the broader management process. But it suffers from a basic logical flaw: it may be sensible for IT people to take a business view, but they also need to have a good grasp of technology – when all's said and done, somebody has to make all the technology work. In a sense, it doesn't matter *where* the line is drawn between the technology and business teams – the point is, the line has to be drawn somewhere. Even if CIOs become business people, either they or their managers are going to have to talk to technical people. Wherever we draw the line between business and technology people, we need to make sure that people can communicate across it.

Perhaps the problem lies with the business people who make up the management team. It is often said that business people need to understand more about computers, and think about them as part of every decision. This is really just the same point as above in reverse: the hope is that by turning business people into technology people, we can somehow improve the level of communication between the two. But very few people have the time or the background to be true experts in both business and technology. So there still needs to be some way of improving their shared understanding of IT projects.

Why Value Estimation Remains Elusive

In summary, we have moved a long way from the replacement of simple business processes towards a sophisticated, networked platform of information systems. Great strides have been made in improving the technologies and the processes of software development and project management. But we are no closer to an understanding of how to make decisions on what information systems to build in the first place, because we have no framework for consistently assessing their benefits in ROI terms.

There are good reasons why it is hard to estimate the benefits of IT systems. The first is the overwhelming complexity of computer systems. Of course, there are plenty of other complex man-made structures in the world, such as ships and buildings and space shuttles; but they do not share one key attribute of software programs, which is that they are easy to replicate. It is this ease of replication that makes computers so troublesome. Once the first ship or building or space shuttle has been built, the incremental cost of building a second or subsequent one is to some extent reduced, because the fixed design and research costs can be spread over more units. But the cost is not reduced to zero – each new ship or building or space shuttle is still very expensive. By contrast, creating a copy of a software program is a trivial task. Naturally, the software vendor also has to recoup the investment cost of writing the software in the first place, but this can be spread over many hundreds or thousands of copies, making the unit cost relatively low. This combination of complexity and ease of replication is one of the reasons that the IT world is so hard to manage: an organisation can buy a huge amount of complexity for very little money. Reading computer magazines regularly, one finds a constant stream of complaints about the lack of reliability of software. Personally, I think we should be amazed that software does not crash far more often than it does, given its unbelievable complexity and low cost of replication.

The second reason why benefits are hard to estimate is that information systems are built in layers, and these layers interact with one another. This makes it difficult to assess investments in information technology, because it is hard to look at any one system in isolation. Again, there are parallels with this problem in other areas of life. The various parts of a car's engine must be designed so that they fit together. The nozzle of a fuel pump in a petrol station must be designed so that it fits into the petrol filler of a car, and there is no doubt a standard governing this, so that a car can confidently be driven between cities and countries in the knowledge that it can still be filled up with petrol. The same applies to computer systems: for a business to invest in software systems, the management need to be confident that software which is designed to run on the Windows operating system will run on any computer with Windows installed. But this adds complexity to the investment equation, because investments in one layer are sometimes necessary to enable capabilities in another.

The third reason why IT investments are difficult is the fact that they are fundamentally concerned with *change*. Software projects are nearly always concerned with changing the way an organisation works. Now, this problem is well understood from the perspective of *implementing* the project. Today, most technology people understand the importance of change management, and most organisations try to ensure that the change process is a smooth one. They may not always succeed, but in general they understand the problem. But in broader terms than this, the fundamental *value* of an information system project derives from changing the way a business works.

The fourth reason why computing benefits are hard to estimate is the essentially unknown nature of the future results of an investment in information systems, the effects of which are very hard to predict. Actually, we might go further and suggest that many of the effects are effectively *impossible* to predict. For example, we might give someone a computer that runs Microsoft's Excel spreadsheet software. But can we really predict what that person is going to do with it? There are people who have built entire accounting systems using Excel, and others who have built financial models of mind-boggling complexity. None of this could have been predicted in advance. The point here is that, contrary to accepted wisdom, it is impossible to anticipate all the benefits of an information technology investment.

The fifth and final reason why IT investments are difficult is that people interact with information systems in unpredictable ways. In the main, computers are used by people, and this is a key factor in the unusual nature of information-system investments. The interaction between computers and people is hard to predict, and even harder to control. This unpredictability is a very relevant factor for businesses making investments in computers. Some systems have no value at all unless everyone contributes. A customer-relationship management (CRM) system, for instance, needs to be fed by everyone concerned; if some people fail to contribute, the organisation will not obtain a complete picture of its customer base. And there are many reasons why people might choose not to participate: lack of interest, lack of relevance, a perception that the system is taking power away from them, and so on. Perhaps the most important is that a CRM system can easily become a system of control rather than a source of information. Instead of simply storing information about a company's clients, the system can be used to set sales targets. Under these circumstances, people will naturally tend to rebel by withholding information.

To summarise, there is no other class of investment or capital project which shares all of the characteristics set out above; complexity and ease of replication, a layered technology structure, the ability to change a business, the unpredictability of benefits, and complex interactions with people. In such a context, the ROI calculation fails to work – the IT world is simply too complex.

Part II
Towards a New Framework

In Part I we reviewed some of the major difficulties with the current methods for assessing IT projects. At the moment, businesses tend to rely on a project-gating approach, based on a core assessment of the ROI of each project, and supplemented by good judgement. But ROI is a poor measurement tool because it fails to capture all the benefits of IT projects, and it seems less than sensible to use it as a yardstick for assessing such projects when it gives results which vary from project to project. So it is hardly surprising that businesses either do not use ROI, or use it only in conjunction with other metrics.

In short, we need a new framework for assessing investments in IT, so that investment decisions can be put on a more solid footing. Perhaps the current way of thinking about IT projects is simply misconceived, and perhaps that is because of a fundamental misunderstanding about what IT systems are really *for*.

What if we were to re-think how information systems interact with an organisation, and how they change over time? Part II presents a three-dimensional framework which connects the worlds of business and IT and shows how they interact over time. Such a framework offers insights about the value of information systems and about the nature of IT strategy.

In later sections, we will see how the framework can be used to assess the value of computer-system projects and to build a rough and ready way of comparing them.

Chapter 3
The Value of Information Systems

If we want to connect the two very different worlds of business and information technology, we need to have simple models for each. Let's begin with the world of business: what model can we use, in representing a company, that might be useful in building a shared understanding between business and IT people?

A model of business

A business can be looked at from a number of different angles: as a team of people, a platform for innovation, a collection of assets, or a legal entity. All of these are valid as descriptions but if we are thinking about a business in order to solve a particular problem – how to assess the value of information systems – it might make sense to think about it as a value-creating entity, a machine for creating value.

A strong candidate for our business model is the value chain, a concept developed by Michael Porter which is widely used and understood in the business world.[9] Figure 1 below shows a standard value chain, appropriate for the large number of companies that buy and process goods, and easily adaptable for other kinds of businesses, including those in the service sector.

Figure 1 – The value chain[10]

Inbound logistics	Operations	Outbound logistics	Marketing and sales	Service	Margin

Porter also shows those functions which support the primary value-adding activities, including technology development, underneath the core value chain. Understanding this fact, that technology development is a support activity, is critical if we want to make better IT investments.

A model of IT

This is one side of the picture, the business side. Is there a similar model that represents the world of information technology? The engineers who build IT systems also use a wide range of models, but there is one simple diagram that is useful for our purpose – the 'technology stack'. Figure 2 shows an example.

Figure 2 – A technology stack

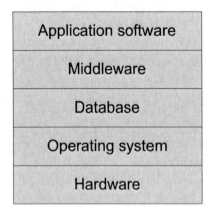

The idea of a technology stack is a useful shortcut into the complexities of IT. The stack is the layered structure of IT systems which together support a business application. For example, when you use Microsoft Office, it runs on Windows, which runs on your PC, which runs on a network. If the layers are well designed, the whole stack functions correctly, while the user is not even aware of it. And a well designed technology stack has some interesting features. First, each layer of the stack has to be able to operate with the layer immediately below it, but not with any lower layers. Second, a lower layer has no 'understanding' of the layers above it. Third, in order to provide the best and widest support to the layers above, each layer must be built to standards that are widely published and understood.

We now have two models: the value chain, representing the world of business, and the technology stack, representing the world of IT. These can be brought together in a very natural and intuitive way, as shown in Figure 3. This simple combination shows how information systems provide support to the business functions in the value chain. And it allows for some useful insights.

Figure 3 – Combining the value chain with the technology stack

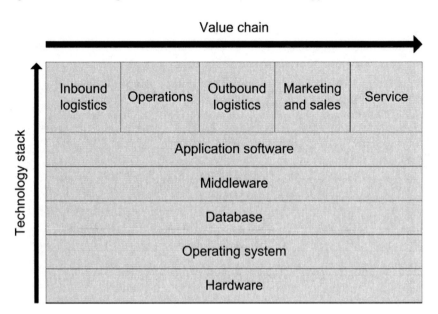

The model implies that it is only the top layer, the value chain, which can directly add value, because it is here that the business actually operates. The highest layer of the technology stack, the application software, can *support* the business, but it cannot add value directly. And the lower technology layers do not even support the business directly, but simply support the layer above them. Contrary to the claims of many information-technology vendors, the lower layers of the technology stack do not *directly* add value to a business.

This simple model also helps to explain some of the complexity of information systems. The picture in Figure 3 above actually shows an idealised technology stack: there is only one software application, one middleware element, one database, one operating system, and so on down the stack. But this is really an idealised picture – the actual situation that most companies face is much more complex. To see this, try to imagine each layer of the technology stack being divided up into systems that support each separate functional area of the business. This is shown in Figure 4.

Figure 4 – A more realistic technology stack

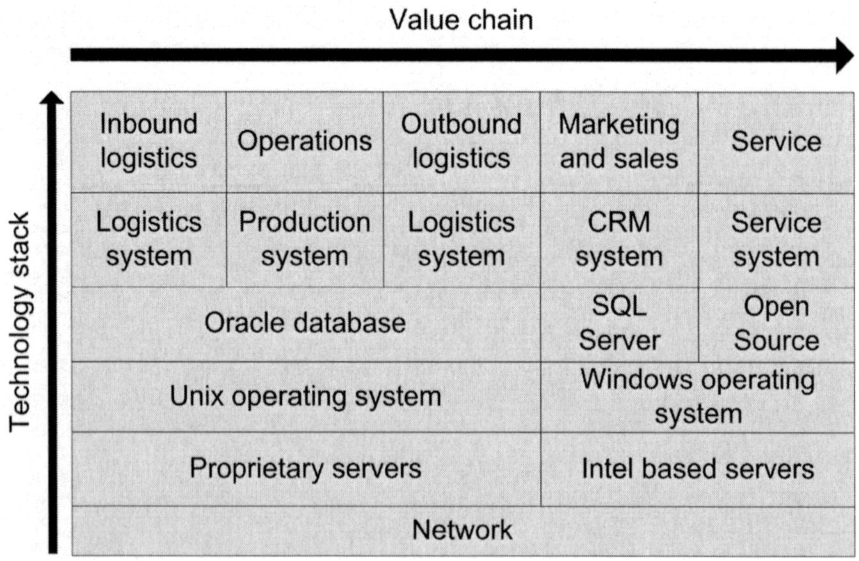

Value chain

Inbound logistics	Operations	Outbound logistics	Marketing and sales	Service
Logistics system	Production system	Logistics system	CRM system	Service system
Oracle database			SQL Server	Open Source
Unix operating system			Windows operating system	
Proprietary servers			Intel based servers	
Network				

(Technology stack — vertical axis label; Value chain — horizontal axis label)

Figure 4 shows the problem that enterprise resource-planning (ERP) systems are supposed to fix. A single ERP system is designed to replace all the applications which support each functional area, providing a common view of the business and a single store of information. In a sense, the first picture of the stack, in Figure 3 above, is an idealised picture of an ERP system, running on just one operating system and one hardware platform. Very few businesses have achieved this level of integration.

Although we can divide the technology stack into many layers, it is convenient to consider just two, at least when deciding on investments in information systems. These two layers are the application software, which impacts directly on the business, and the technology infrastructure, which is there to support application software. The technology-infrastructure layer could be defined as 'any software or hardware component that does not directly interact with a company's people or processes'. This is shown in Figure 5.

Figure 5 – The value chain and simplified technology stack

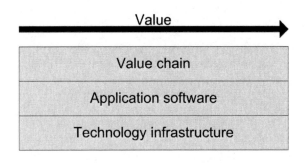

This clean split between applications and infrastructure is a natural one, and one that a lot of IT departments already use, with their teams being divided between the two. Of course, there is no perfect dividing line between the two technology layers. For example, email is a software application according to the definition above, but it is effectively treated as technology infrastructure by IT people. The argument here is that this is the exception rather than the rule. For most businesses, most of the time, the technology infrastructure is there simply to support software applications.

Distinguishing between application software and technology infrastructure

This cuts to the very heart of the problem. According to our new model, an investment in a firm's technology infrastructure cannot possibly show any benefit at the business layer, because it is two levels down and has no direct impact on the creation of value. After all, a server is a server, whether supporting a manufacturing plant or a bank. Yet the implicit assumption behind an ROI calculation is that an investment in technology infrastructure must show benefits at the business layer, which is two levels up! Of course, an investment in the technology infrastructure *can* show a positive ROI, if the initial investment yields longer-term cost reductions. But this is *not* helping the firm to add value at the business level: after the investment has been made, the new technology infrastructure is doing exactly the same job as before, albeit slightly more cheaply.

Looking ahead, it is clear that there should be different business cases made for IT projects, depending on whether the latter are concerned with application software or with technology infrastructure. An investment in application software *might* help a business to add more value, but an investment in technology infrastructure will definitely not do so, since the technology infrastructure has no direct effect on a business.

An interesting and controversial paper by Nicholas Carr, appropriately titled 'IT Doesn't Matter', suggested that computer systems are a commodity which cannot give a business any competitive advantage. Carr compared IT to the electricity grid, observing that everyone is connected to the grid, that electricity benefits everyone equally, and that no business gains a competitive advantage from it.[11] But his argument about information technology in general is really applicable to the technology infrastructure alone; this *can* effectively be treated as a commodity, because it can be freely purchased and confers no competitive advantage. By contrast, the application-software layer interacts directly with a business's people and processes. These are constantly changing under competitive pressure, and the interaction gives a business the potential to gain competitive advantage.[12]

Many businesses are already applying this simple distinction. The third largest business in the world, ranked by market value, is General Electric Corporation (GE).[13] If any business is capable of managing its own technology infrastructure, it is GE, with its massive size and unparalleled expertise. But when the company wanted to rationalise and improve its global procurement system, the CIO, Gary Reiner, decided to implement a system that is hosted and managed externally. In an interview with *CIO* magazine, he said "When we judge a solution, we are indifferent to whether it's hosted by a supplier or by us. We look for the functionality of the solution and at the price."[14] In other words, GE understands that technology infrastructure is a commodity, but that the functionality built into the application software is not.

Implications for project assessment

I have suggested that information systems have only a supporting role in most businesses, and cannot directly add value. If that is true, there are some important implications for project assessment. Let's return to the example of a taxi business. Now, there is one big difference between buying a new taxi and buying a new accounting system: the taxi is an investment at the business level, while the accounting system is an investment at the application-software level. This suggests that we need a different approach to evaluating such investments. And when we go down one level further, to the technology infrastructure, things become even more complicated. When the taxi firm invests in IT infrastructure, such as new servers or an improved network, the effect can only be felt at the application-software level. These three different classes of investment project are shown in Figure 6. Again, one immediate conclusion is that it is wrong to focus on ROI as a yardstick for all projects. The ROI approach is appropriate for business-level investments like a new taxi, but it fails to work for technology-level investments that support the business – except insofar as they may reduce direct IT costs.

Figure 6 – Different types of investment in a taxi business

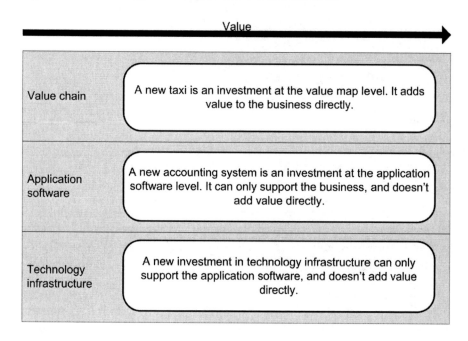

The Benefits of Applications

If applications cannot add value directly, what exactly *can* they do? As a very broad generalisation, application software offers benefits to a business in two key ways: automation and information. Automation can improve the efficiency of a process, and can also improve the productivity of a person or group of people. Information can improve communication between people and processes. These two kinds of benefit provide a link between the value-chain and application-software layers, as shown in Figure 7.

Figure 7 – Automation and information benefits

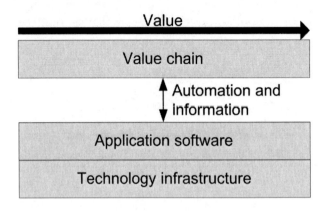

Automation

Automation improves the productivity of people and the efficiency of processes. It helps businesses to deliver more or better goods or services more quickly for less input cost, and IT is an important enabler of such process improvements.

For example, Cemex is the largest supplier of concrete in Mexico, and a major player worldwide. Although not perhaps a glamorous business, concrete supply is certainly a time-critical one. Concrete needs to be used before it sets, and the giant trucks that transport it to construction sites need constantly to rotate the drums they carry in order to keep the concrete liquid. In the 1980s, Cemex was able to guarantee delivery to construction sites only within a three-hour window. In the worst-case scenario, that meant a construction crew would have to wait three hours for their concrete. So Cemex decided to implement a routing and scheduling system that connected to computer terminals and satellite navigation devices in each truck. By better management of its fleet of trucks, Cemex was able to reduce the delivery window from three hours to 20 minutes. In doing so, it was able to charge construction companies a premium price, since their crews would spend less time waiting for a Cemex truck than for a competitor's. Operating costs, including fuel, were also reduced, when the number of trucks required to service Cemex's customers went down by a third. This is a classic case of the efficiencies that can be gained through automation, and the benefits that in this case were shared between Cemex and its customers.[15]

Information

What about information? Systems like intranets, email and wiki pages are aimed at sharing information, providing a communications platform within the business. But information systems are also used as control mechanisms. The budgeting process within a firm is one of the key means by which the management team exerts control – computer-generated information tells managers what is going on and how much it costs. And information also allows managers to understand and control their business better, by finding and fixing problems more quickly. Running a business with poor information is like flying blind.

In 2004 the leisure group Whitbread plc became the largest hotelier in the United Kingdom when it bought 144 Premier Lodge hotels to add to the Travel Inns it already owned. After the acquisition, Whitbread standardised the booking and billing systems across all its hotels, as part of a large integration project. This allowed the firm to increase the volume of cross-selling across its hotels. Now, when a customer's preferred hotel is full, there is a higher probability that Whitbread will be able to provide a room in another hotel nearby. The group made £16 million between 2004 and 2005 from such cross-selling. Of course, Whitbread may also have reaped benefits in terms of IT support costs, but the real benefits derived from the improved information flows between hotels and head office.[16]

The importance of people and processes

In some cases, it is easy to distinguish between automation and information benefits. For example, a call-centre system gives benefits largely through automation – calls are routed more efficiently with less staff time. Conversely, an intranet gives benefits through improved information, by making knowledge available more quickly to more people. But most of the time, it is not possible to ascribe all the benefits of a system to either automation or information, because the two effects are mixed. An ERP system gives automation benefits, and that is often the primary reason for implementing one, and it also provides information, by giving a business a single view of its data. But this ambiguity does not matter so long as the real benefits of a system are clearly understood.

The key point to remember about automation and information is that *they are connected to people and processes*. These connections are not easy to build – it is hard to configure a process to work with an application. It takes time and is prone to error (getting lost when trying to navigate through an automated call-centre system is a common experience). It is equally hard for people to learn a new software package. There is a kind of entropy effect whereby they forget about software features, and at the same time information requirements are evolving and changing as the business moves on. Application software adds value via some very direct but complex interactions with people and processes.

When Kimberley-Clark installed the SAP ERP system across its 32 manufacturing mills, the project took three years and cost more than $100 million. A large component of that cost was the need to train front-line mill workers on the new system. Many were unused to modern business software, and were unhappy at its introduction. According to an interview with Kay Chase, a senior IT analyst, in *InformationWeek* magazine, "This was an aggressive undertaking, because a lot of people in the mills did not use computers. We were starting from scratch on how to use a mouse to how to work in the SAP system." It took time and effort to make the links between people, process and technology, but Kimberley-Clark persisted, and planned to roll out the new SAP-enabled processes outside North America.[17]

The Benefits of Technology Infrastructure

If improved automation and information are the means by which applications give benefit to a business, what does the technology infrastructure layer do?

Again, as the framework indicates, the technology infrastructure does not add business value directly, but instead supports software applications. So it makes little sense to assess the business benefits of technology-infrastructure projects directly, using ROI, because the infrastructure is not directly connected to people and processes. In fact it connects to the application-software layer via standards-based interfaces. These two very different kind of links are shown in Figure 8.

Figure 8 – Interfaces to the technology infrastructure

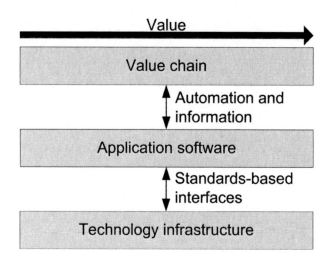

For example, Windows applications are connected to the Windows operating system via an application-programming interface (API) called the Windows API.[18] This helps to ensure that all Windows applications can run on all Windows PCs. These standards-based interfaces make it possible to change elements of the technology stack with some confidence that the resulting solution will be viable. This is not to say the work is easy; but the problems are more predictable than those which occur at the application level, when people and processes are involved. At the very least, the projects should be well defined.

However, the very fact that these interfaces are standard and are published (in this case by Microsoft) means that they are unlikely to be a source of competitive advantage. Again, Nicholas Carr had it half right: technology infrastructure cannot help a company build a competitive edge. But application software, configured to support people and processes, can.

Chapter 4
Organisational Complexity

L et's go back to GE, a global organisation with revenues of $183 billion and over 300,000 employees. It is spread widely, generating revenues above $1 billion in each of 25 countries around the world. It has businesses that make aero engines and gas turbines, provide capital finance, harness wind and solar energy, or operate in the media industry and in healthcare.[19] Given the overwhelming size and scale of GE, how can one even start to plan the information systems? There is a fundamental problem with planning IT investments without taking into account complexity.

A value chain is an abstract representation of a business. While useful as a conceptual model, it offers little help in tackling the complexities of a corporation the size of GE, or even of one of GE's operating businesses. Most large companies consist of more than one business unit, and even a small business might have two value chains rather than just one.

To add complexity, many companies are spread over large geographical areas. Even in today's world of globalised trade and finance, geography still places constraints on businesses, and this applies equally to information systems; even with fast data networks, geography still affects the way in which software is architected and deployed. Perhaps this will change with the current shift to 'cloud computing', and perhaps businesses will be able to rent software from anywhere in the world whenever they need it. But for the moment, geography still matters.

Further complexity comes from organisational structure. Even if a company operates in just one geographical area, it might consist of multiple divisions or operating businesses. Groups which have grown through acquisition often have a highly complex organisational landscape.

As an example of a large and complex group, take the diamond company De Beers, which has revenues of $7 billion. According to the company, "Group Exploration is focused on projects in Angola, Botswana, Canada, the Democratic Republic of Congo, India, Namibia and South Africa", while "De Beers mines for diamonds in Botswana, Namibia, Canada and South Africa", and has sales and distribution centres all over the world.[20] A simplified view of De Beers's worldwide operations is shown in Figure 9.

Figure 9 – The worldwide operations of De Beers[21]

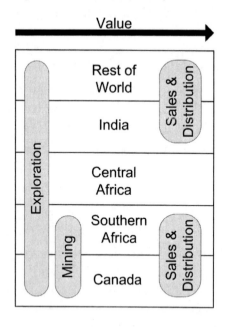

Complexity really matters for IT systems. Any framework which fails to take account of it is not very useful for making IT decisions (and this is certainly true of ROI). We need to combine the abstraction of the value chain with the complexities of geography and organisational structure. We can do this by adding complexity as a second dimension to the simple value framework. This creates a slightly more complicated picture, as shown in Figure 10 below, where the operations for De Beers are shown in the business layer at the top.

Figure 10 – Introducing complexity into the framework

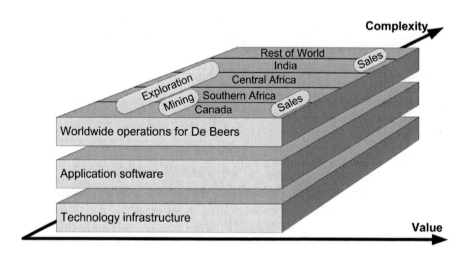

We have now extended the simple value chain and technology stack into a two-dimensional 'sandwich' of three layers, representing the real-world entities that make up a business. For the sake of simplicity, this can be termed a 'value map'[22] – a map of the way a business adds value, and how technology supports it.

While assigning names, we may also refer to the business layer as the 'operating model', which consists of the people and processes that make the business work. Below the operating model is the application-software layer, which is the set of applications that support the business's people and processes. The bottom layer is the technology infrastructure, which supports the application software. The operating model and its supporting layers are shown in Figure 11.

Figure 11 – The operating model and the value map

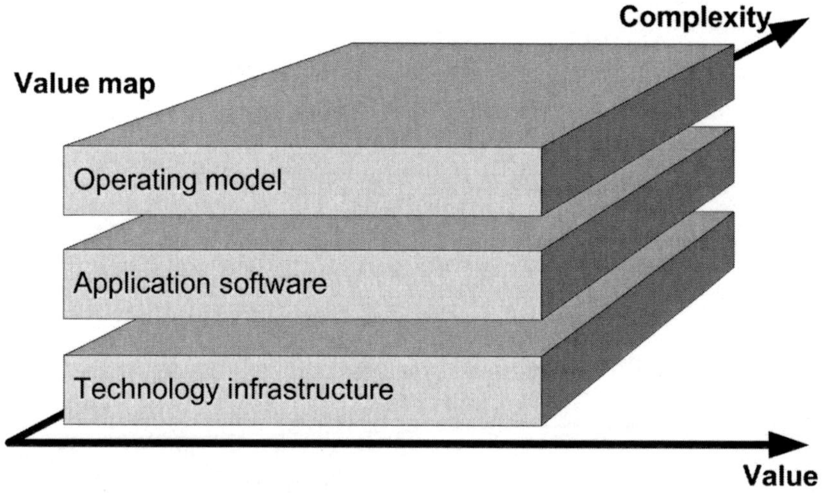

Although the framework shows value along the first axis and complexity along the second, this need not be rigidly applied. The operating model is simply a two-dimensional diagram that shows value and complexity at the same time. It is the kind of rich picture a CEO might draw in order to show how the business works, a picture of the people and processes within a firm that combine to create value. This framework adds an important element to companies' internal debates about the value of IT projects: it allows managers to include complexity explicitly within their decisions.

However, the value map is proposed here as a conceptual framework rather than as a practical tool for mapping systems. After all, the value map for a group the size of GE would be a highly complex diagram, and there would be little point in producing it for its own sake. The framework is intended to make decision-making more consistent and to try to encourage a dialogue between the business and IT teams. The scale and granularity of the value map should be adjusted to suit the context.

There is a nice analogy with real, geographical maps. There is no single 'right' map for every situation: if we decide to visit friends in another country, we might start with an atlas of the world, to see roughly where they live and to plan flights. On arrival in their country, we might switch to a large-scale road map to navigate the way to their home town. Finally, we might pick up a local street map to find their house. In other words, we use a variety of maps depending on our purpose. In the same way, there is no single, right value map for a business – the value map is not intended as an all-embracing picture of a company, but rather as a useful tool for making decisions about IT investments.

Different insights can be gained from different scales, and this allows for debates within a management team. At a detailed level, for example within a single plant, the value map might represent specific processes and people. This level is conceptually close to that adopted during the normal requirements-gathering and software-specification processes which IT people currently use, as shown in Figure 12.

Figure 12 – Example of a small-scale value map

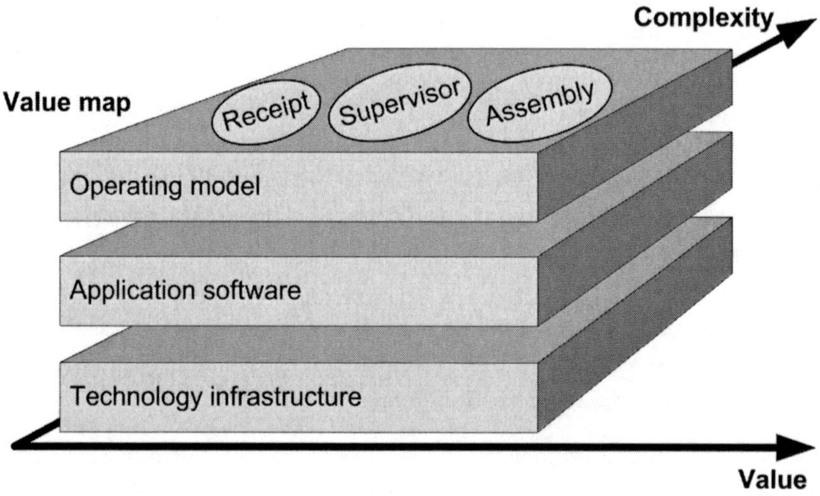

Identifying real-world improvements

It is only at this detailed level that opportunities for real-world improvements can be identified with any precision. As an aside, this has important implications for IT value: investment decisions must be made at the top level of business, because of the amounts of money involved, yet value is gained at the front line, inch by inch, day by day, by automating processes and giving people better information. For that reason, it is difficult to identify all the potential benefits of a new IT system in advance. When managers agree to go ahead with a new system, they do not really *know* what the benefits will be, because these benefits must be won at the front line.

Moving up in scale, the value map can be imagined for a whole group, treating each component business as a separate entity, as shown in Figure 13. At this high level, it is no longer feasible to think about particular processes and key people, but other insights can be gained. In particular, it is possible to visualise and debate the benefits that might be derived by sharing information, or the technology infrastructure, across the group.

Figure 13 – Example of a high-level operating model

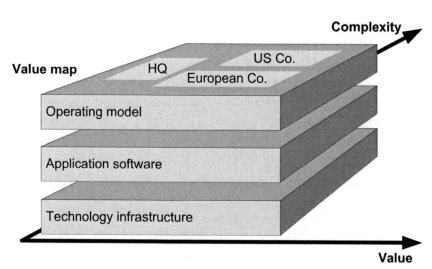

The importance of people

The operating model shows people and processes at the same level – neither is subordinate to the other. The fact that people and processes are given equal importance is a deliberate rejection of the commonly held view that all business functions should be reduced to processes. The process-centric perspective was clearly set out by Michael Hammer in his book *Beyond Reengineering*:

> *Processes must be at the heart, rather than the periphery, of companies' organization and management. They must influence structure and systems. They must shape how people think and the attitudes they have.*[23]

In this perspective, processes are the core unit of analysis when businesses examine their operating models. And it is by improving processes that businesses will approach the world-class standards needed to compete.

This process-centric view of the world has gained in popularity, and there is no question that it can be effective; but processes are not always or necessarily more important than people – indeed, the latter often play a much bigger role than processes in creating value. For example, law firms rely heavily on the knowledge and skills of their lawyers – they are people-centric businesses. Although it is possible to think of a law firm as a set of processes, the result is less than useful. Imagine a lawyer in court, passionately arguing the case for his or her client; can that really be reduced to a 'court-room process'? It feels like an attempt to squeeze reality into a theory, rather than to design the theory to fit the reality, and it is surely better to accept that some sorts of firm are inherently people-centric.

Furthermore, by putting people at the same level as processes it is possible to make better decisions on IT priorities. From an IT perspective, a top lawyer may use nothing more than standard productivity applications, such as email and Microsoft Office. But from a *business* perspective, the lawyer's work is critical. That insight may help a company to assign the right priority to the lawyer's IT projects; a simple request to design a new spreadsheet could be seen as a critical project, if the lawyer needs it urgently.

Harris Corporation is a $5 billion communications and IT group with 15,000 employees, including 6,500 engineers and scientists.[24] When Harris needed to improve the performance of its supply chain, it could have opted for a process-centric approach. But the company decided instead to focus on its engineers and the purchasing decisions that they had to make, by building a new data hub that provided them with detailed information about the costs of the components they used when designing new products. In an interview with *Information Week* magazine, Harris's director of supply-chain management, Janice Lindsay, acknowledged the key role played by the engineers when she said: "The engineers control our supply-chain destiny."[25]

This people-centric approach also highlights the importance of productivity tools, where people use software directly to enhance the way they work. Once again, the process is not really the central focus. For example, a firm of architects might use a computer-aided design (CAD) software package that allows them to design buildings more quickly, to change designs and test ideas, and to exchange design information with people in other locations. But this is not really a *process* – essentially, the architects are talented people who work with software tools, and the process is irrelevant. Flower and Samios, a small architectural firm in Australia, started to use CAD software only after they had been beaten in a selective bid for a large project. But having accepted the need for automation, the principals were enthusiastic and committed to the benefits that would thus be obtained. And staff members agreed, with some saying: "I thought it was quicker to do manual drawings, but, after a few weeks, I was much faster on the computer."[26] The value of the CAD system was to be found not in processes, but in making individual architects more productive.

Again, by placing the information systems underneath the people and processes in the business layer, we are saying something fundamental about the relationship between the two: that the primary role of IT systems is to support a company's operations and to improve their efficiency. It is the company's operating model that sets out its approach to serving its customers, and thereby generating a profit, and software applications are only there to support it. Does this mean Nicholas Carr was right – that information systems are simply a commodity? Not really: a company can gain a business advantage from computers without elevating them to the status of strategic enablers.

Swisscom is Switzerland's largest telecommunications company. In 2008, the firm gave each of its field engineers a ruggedised PC running specialised scheduling and dispatching software. This allowed for much more accurate scheduling of the engineers' time, and led to a 10 per cent increase in their productivity. So the system gave Swisscom a business advantage, but was hardly a strategic enabler; the advantage was incremental, and there is obviously nothing to stop the firm's competitors from making similar improvements.[27]

It might be felt that this view of IT's role as simply a supporting platform is somewhat limiting, but a more compelling view is perhaps that the framework puts computers in their proper place. The management expert Peter Drucker observed: "Because its purpose is to create a customer, the business enterprise has two – and only these two – basic functions: marketing and innovation. Marketing and innovation produce results; all the rest are 'costs'."[28]

Only human beings can think creatively about how to do things differently, and how to serve their markets better. IT can automate many of the tedious and repetitive tasks that people would otherwise have to perform, while also giving them better information and thus freeing them to focus on the core creative functions that Drucker identified, namely marketing and innovation. Of course, this is a simplification – people are rarely short of boring and repetitive tasks. It would be more accurate to say that there is always commercial pressure on companies; that this translates into pressure to find more productive ways for people to work, since salaries are generally a company's largest cost; and that this tends to mean they carry out fewer and fewer of the boring repetitive tasks that can be better done on a computer.

Actually, computers can help to *support* the key activities of marketing and innovation, even if human beings are the key to both. Procter & Gamble (P&G) is the world's largest consumer-goods company. But scale alone will not guarantee P&G's number-one position: in order to stay ahead, the group needs constantly to innovate, finding new ways to make people's everyday lives a little bit easier and a little bit better. As A.G. Lafley, P&G's Chairman and CEO, said in his book *Game Changer* (written with Ram Charan):

> ...*we knew that innovation would be the key to winning over the medium and long term. Why? Fundamentally, P&G had been built on a strategy of differentiation – of differentiated, branded consumer household and personal-care products.*[29]

P&G relies primarily on its research scientists to deliver a continuous stream of differentiated product innovations, and the company made an interesting decision to support this process with a computer system. The concept of product-lifecycle management software is still a relatively new one, but P&G has used it to try and ensure that the pipeline of new product ideas is full, and that scientists in one part of its vast worldwide organisation are not duplicating work being done in another. In a pilot programme, researchers were encouraged to record their laboratory notes in a common repository. This allowed their peers across the world to access their notes, look for common ground, and perhaps share ideas. By encouraging researchers to share a common repository, P&G has attempted to maximise the efficiency of the whole process. It is still people who do the real innovation, but information systems can at least support them.[30]

None of this means that software is insignificant. For example, Amazon Corporation's online ordering process relies on a sophisticated software platform, and this software is visible to the customer (in fact the software itself is now being made available to third parties over the internet).[31] But even so, it is arguable that Amazon's customers are really responding to the *effects* of the software rather than to the systems themselves. People buy from Amazon not because they love its computer systems, but because they can find the books they want, the delivery times are reasonable and the prices are fair. For most businesses, the application software is there to support the firm's key value-adding activities, as expressed in the operating model.

Chapter 5
The Change Dimension

Many people and organisations are stuck in a cost-cutting mindset when they think about IT investments. Whether explicitly or implicitly, it is assumed that the IT systems should simply support the current operating model, as cheaply and reliably as possible. Businesses spend a lot of time trying to drive down the cost of the IT provisions, squeezing the same level of IT support from a lower cost base. But this is the heart of the problem with today's approach to IT investment – a company's information systems cannot be optimised simply by cost-cutting. Why is this? What is wrong with the simple two-dimensional value map?

The answer is that the picture is a *static* one. It shows business as usual, and encourages a perspective that cost-cutting is the best that can be done with the IT platform. And this means it fails to engage with some of the most fundamental characteristics of information technology.

First, software does not decay. Once a computer program is written it will never change. This lack of decay makes software utterly different to almost every other complex system. Bridges rust and collapse, roads break up, cars break down, and paintings fade. But software programs are probably the only complex man-made systems that do not decay. This point is often missed when discussing software maintenance – we unconsciously assume that software must deteriorate over time, and we talk about software getting old and breaking down; but this does not happen in reality. Software never simply breaks without human intervention (though it is easy enough to break a computer program by modifying it, even when trying to improve it).

Secondly, information technology is itself an agent of change within firms. When a business process is automated, it leads to changes in processes and people. So it is not sufficient to see IT simply as a static entity supporting the current operating model, because it also drives change.

Thirdly, changes in IT often lag behind changes in people and processes, because the technological elements are so complex. There will then be a mismatch between the current requirements of the business and the capabilities of the application software. As business processes evolve, the IT applications that underpin them can lag behind, simply because of the time it takes to develop and configure them.

Fourthly, the technology itself is always changing, so that a static view of IT is generally going to be supplanted by new versions of the underlying technologies. The normal, static view of IT is insufficient to cope with this dynamism.

Finally, software is about development rather than production. When programmers write a software program, they do not really know how the final product will turn out. Even with the best design and specification process in the world, there will be variations. Information technology is really a developmental activity, not an operational one, more akin to the development of a new product than to its production.

For all these reasons, an IT system is more than just a static support function, but also acts as an enabler of change: in this it is fundamentally different to other support functions. In an interview with *CIO* magazine, John Hill, the CIO of Praxair Corporation, a $5-billion producer of industrial gases and surface coatings, put it very well:

> *Managing IT is really managing two kinds of functions. One is a utility, which is cost-driven and focused on service levels. The other is like a venture capital fund where you manage resources with the idea of maximizing returns.*[32]

The change dimension

Most businesses zero in on the first function: they spend their energies in ensuring that their IT teams support the business and adhere to service-level agreements. But sooner or later this narrow focus will cause a firm to make poor IT decisions. We need to broaden the debate within firms on the real value of their IT projects by including the second function, where IT acts as a driver of innovation. The importance of change can be reflected in the framework by introducing a third element, the 'change dimension', as shown in Figure 14 below. The framework now explicitly incorporates the concept of change, starting with the current value map and moving to the future.

Figure 14 – The change dimension

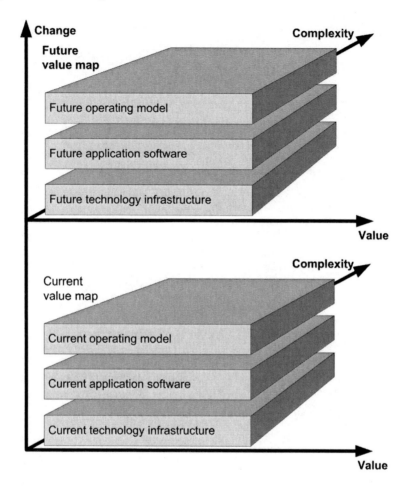

How do we get from the current value map to the future one? Let's start with the business layer at the top of the current version. There will need to be a set of business projects that transform the current operating model to its future state. The projects can encompass anything from new acquisitions to operational improvements. Changes in the operating model might allow a firm to respond to changing competitive pressures or to build a new strategy that leapfrogs the competition. In other words, the changes that are planned for the people and processes of the business are all conceptually mapped onto the new future state of the operating model. This is shown in Figure 15.

Figure 15 – Business projects in the change dimension

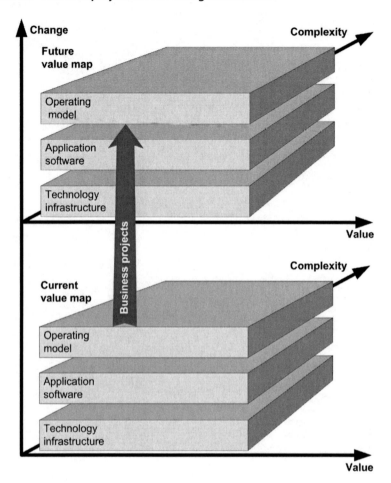

Just two pages from a single issue of *The Economist* covered news of several business projects. First, an American company called Cablevision was rolling out interactive advertisements to its cable customers. Next, two mobile-phone businesses in Russia, jointly owned by Alfa Group of Russia and Telenor of Norway were to be merged. Finally, Carrefour, the French supermarket giant, was considering selling its Asian and Latin American businesses. All these projects can be thought of as changes to the respective companies' operating models, and all would be bundled into the arrow marked 'Business projects' in Figure 15.[33]

There will also be projects for the two lower layers in the value map, the software applications and technology infrastructure. Taken together, these three parallel sets of projects will move the three layers of the value map to their corresponding future states, as shown in Figure 16.

Figure 16 – Three sets of projects in the change dimension

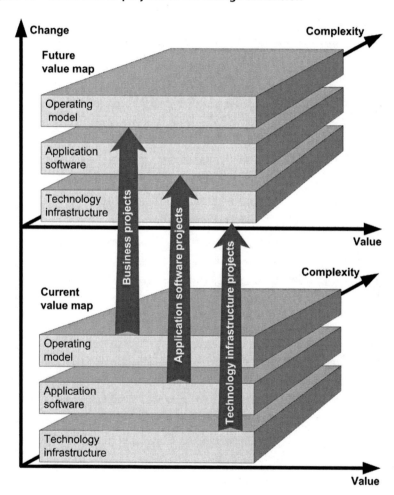

Flexibility and future development

At this point, it is tempting to think about alignment between the business and the technology plans. In the diagram above, alignment would seem to imply that the three sets of projects are mutually supportive and closely synchronised. This is the implicit assumption behind the current vogue for alignment between business and IT strategies. Such a conclusion is tempting, but flawed: there is a missing element – *flexibility*. The flexibility of each layer of the value map gives *options* for the future development of the layer above.

For example, the installation of an ERP system will increase the flexibility of a company's operating model by enabling a large number of options for the future. A business that installs an ERP system will not generally attempt to absorb all the corresponding changes in people and processes at once. Many of the features that the ERP software provides will be unused when the system first goes live; they will represent options that the business can take up in the future.

The consumer-products group Clorox implemented SAP's ERP system in exactly this way. In 2003, it completed the roll-out of the basic components of SAP, including customer orders, accounts receivable and customer service. The next step was to install SAP in its manufacturing operations, and to link its marketing promotions with production. As the final stage in the phased implementation, Clorox planned to install the human resources, supply chain and financials for internal operations.[34]

Typically, the finance system is one of the first elements of an ERP system to be implemented. Until that system is running, there can be no common approach to cost-accounting across the business. But modules such as human resources and materials planning are not implemented until later: they represent future options that the business can take up.

Farwest Steel, a steel maker based in Oregon, explicitly built this flexible thinking into its decisions on IT applications. When the management team decided to replace their ageing legacy system with Oracle's modern E-Business Suite, they decided to break the implementation into two phases. The first phase, 'modernisation', was aimed at replacing the legacy system and getting a new platform up and running. At this stage, the core ERP modules were implemented, including finance, sales, procurement, production and inventory control. The second phase, 'extension', has the objective of building better collaboration capabilities, thereby enabling improved customer service and faster decision-making. During this phase, Farwest plans to build on the core ERP platform, adding such features as supplier collaboration, warehouse management, customer-relationship management, sales-force tracking and improved e-business capability. In other words, the company understood that the investment in Oracle's ERP software was a long-term investment and that the full value would be obtained only through exploiting a series of linked options.[35]

Critically, some of the options will not even be known at the time the application software is implemented. To continue with the example of ERP, no company that installs such a complex system will ever be able to predict all the ways in which it will change the operating model.

Some companies are aware of this, and explicitly build it into their thinking. JetBlue, for instance, is a successful budget airline. Having selected SAP as its ERP system, the company adopted a phased approach to implementation. In the first phase, it planned to install the human-resources and self-service elements of SAP, and in the second, purchasing and materials management, just as Clorox and Farwest Steel did. But beyond that, JetBlue was very open about where the SAP system would be developed next, with no definite plans. A report by *Infoworld* said that: "In early 2006, JetBlue plans to phase in purchasing- and materials-management software from SAP. Beyond that, JetBlue will see what needs arise as it decides how much of SAP's suite to implement."[36] Such an open-ended approach clearly shows that JetBlue is a company that understands the flexible and options-based nature of software applications.

The lack of full knowledge about how software will be used is not a failure of the specification process. No matter how good the process, no business can predict in advance everything that it wants from a complex software application. Most of the changes to the operating model will emerge later as people start to use the new software. The fact is, it is impossible to know how those requirements will change in the future, because software is interwoven with the way people work. By using the software, they will evolve new and unforeseen ways of working. Over time, they will use the system in ways that nobody predicted when it was put in.

The insurance company Assurant found that its website was generating increasing sales in one particular area, that of renters. When a property is rented, the property-management company will often insist that new tenants take out insurance, and the easiest way for them to do this is through the online website. So Assurant is working closely with property-management companies to offer tenants a high-quality service. The decision to approve the insurance proposal can be made very quickly through the online process, much more quickly than the alternative of telephone queries and paper applications. Everyone benefits: the tenant, the property-management company and Assurant itself. And the key is the complex interaction between people, process and application software.[37]

This might make more sense if we think about the numbers of people involved. The implementation phase for an ERP system will typically involve a project team of a few dozen people. But once the system is up and running, there might be thousands of people using the software across the world. Instead of a few dozen people trying to work out how the software *might* be used, there are thousands of people saying how it *can* and *will* be used.

Software value is found in the changes it enables

In summary, the value of an application-software project should be seen in terms of the *changes* which it enables a firm to make to its operating model.[38] These should be evaluated in terms of immediate changes and future options, most of which will not be known in advance. The three-dimensional framework needs to be extended to cope with options, so that a constructive debate can be had within the management team. We start with the business layer, as shown in Figure 17.

Figure 17 – Striking options at the business level

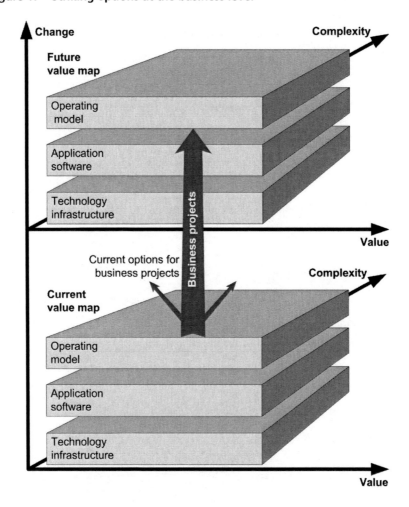

At any given time, the business can optionally strike a range of options enabled by the current application software. This might sound like a theoretical point, but it is not meant to be – it simply states the obvious: that managers can make decisions to change their business's operating model, and that some of these changes will be 'allowed' by the current application software.

Although one of the objectives for application-software projects is to enable immediate changes to a firm's operating model, the benefits of a system for a business often do not come about directly. Rather, software *increases a firm's options* to adopt new ways of automating business processes. This is shown in Figure 18. As the diagram shows, changes in the application software can increase the number and range of future options for improving people and processes.

Figure 18 – Building options from application software

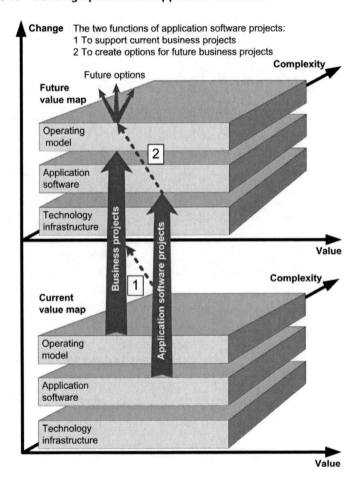

With annual revenues of $52 billion, and 425,000 employees worldwide, the express carrier and parcel-delivery company United Parcel Service (UPS) is much bigger than its rivals.[39] But historically it had failed to emphasise investment in IT, and by 1986 had fallen behind in that area. Over the next ten years, UPS invested heavily in order to catch up. A key component of this was a bespoke system that allowed the company to track the 16 million packages which it delivers each day.[40] Now the package-tracking system has become a platform that allows it to add new functions to meet new business needs. One of these is a 'package flow' application which allows the company to optimise its delivery patterns, so that its truck drivers spend less time waiting at traffic lights and travel less distance to deliver a package. This is an option which probably could not have been predicted at the time the tracking system was developed. As UPS's CIO David Barnes put it: "Package tracking is a great example of how we developed a common technology foundation and built upon it to improve services for our customers."[41]

Similarly, the M.D. Anderson Cancer Center at the University of Texas decided to build a system for electronic records management (ERM). Having built the system, the organisation has found that it is flexible in adapting to new options. It is now the physicians who are driving the development of the system, without being limited by the technology. In fact, the ERM system has given the organisation many more options at the organisational level.[42]

Even though the effects are delayed and unpredictable, businesses still choose to invest substantial sums in information technology. Why? Unless management teams are simply making the wrong decisions, these investments must be generating value. The answer is that the potential gains are high, and can make the delays and uncertainty in the outcomes worthwhile. Investments in information systems are like a long-term bet – the outcome is unpredictable, but the benefits may well outweigh the costs. This is not a flippant comment, nor is it intended to diminish the importance of the subject. On the contrary, the whole question of IT investment is important, but it inherently involves a large element of the unknown. This means that the traditional tools for judging investments are not appropriate, and that a probabilistic approach is much more useful.

Also, it should not be forgotten that while the benefits of IT systems might be hard to measure, they can be startlingly high. A classic example is American Airlines's development of the Sabre computerised reservation system. This was stunningly successful, and allowed American to redefine the way in which the industry made bookings, with other airlines being more or less obliged to use Sabre. At one point, American was making more money from Sabre than from flying passengers. In a congressional hearing, American's then CEO Robert Crandall said: "If forced to break up American, I might just sell the airline and keep the reservations system."[43]

The same thinking applies to the technology infrastructure, in which flexibility gives options for the application software. For example, by having an up-to-date operating system a business gives itself the choice of a wider range of application-software packages. This is shown in Figure 19.

Figure 19 – Building options through technology infrastructure

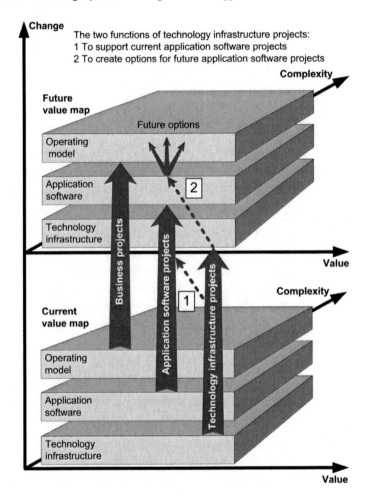

This is the real reason why businesses continue to invest in their technology infrastructures: to support the latest application software. The investments are required to give the business *options* in the future. This is not, however, the reason normally given (businesses might be making the right decision for the wrong reasons). The reason for upgrading an old operating system is not *just* because the software is no longer supported by the vendor: it is because a newer operating system will increase the company's range of options for application software in the future.

For example, Lufthansa has invested in a layer of technology middleware based on services-oriented architecture (SOA). This investment will not create immediate business value. Instead, Lufthansa plans to reduce the cost and complexity of rolling out future changes to its applications, and in particular those applications that connect to Amadeus or other service providers. In an interview with *InformationWeek* magazine, Lufthansa's CIO, Christop Ganswindt, said: "We have a lot of systems that need to communicate with Amadeus. Adding a layer of (SOA) middleware to handle that communication gives us a lot of flexibility." In effect, Lufthansa is investing in flexibility and future options.[44]

This comes back to the point made earlier, that technology can sometimes lead the business, and act as a change agent. This is what happens when a firm invests in new technology capabilities that are not put to immediate use. The technology is ahead of the business, and leading it in new directions. In our framework, we show these as options created for the future.

This discussion has been about value for individual businesses. But there is evidence that similar ideas can be applied at the macro level of the economy as a whole. Erik Brynjolfsson and Adam Saunders examined this issue in their book *Wired for Innovation: How Information Technology Is Reshaping the Economy*. They looked at data from an earlier paper[45] where the authors showed that:

> [C]omplementary investments to IT can take years to come to fruition. Using data from about 500 large firms, they found that the one-year returns to IT were normal, just like ordinary (non-IT) capital. However, they also found that over a longer period (5–7 years) the productivity and output contributions of the same technology investments were up to five times as large. They concluded that the dramatic difference in returns was due to the time it took for the complementary investments in human capital and in business-process reorganization to pay off.[46]

The addition of options is the final piece in our new framework. Now that we have allowed for future options, we can see the final version of the framework, as shown in Figure 20.

Figure 20 – The full decision framework

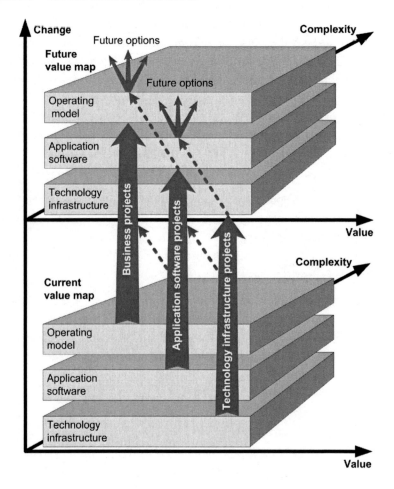

Chapter 6
The Context of Competitive Pressure

L et's round out the new framework by adding two key external elements: investors and customers. These are shown in Figure 21 to the left and right of the framework respectively (with the change dimension omitted for simplicity).

Figure 21 – Adding investors and customers outside the framework

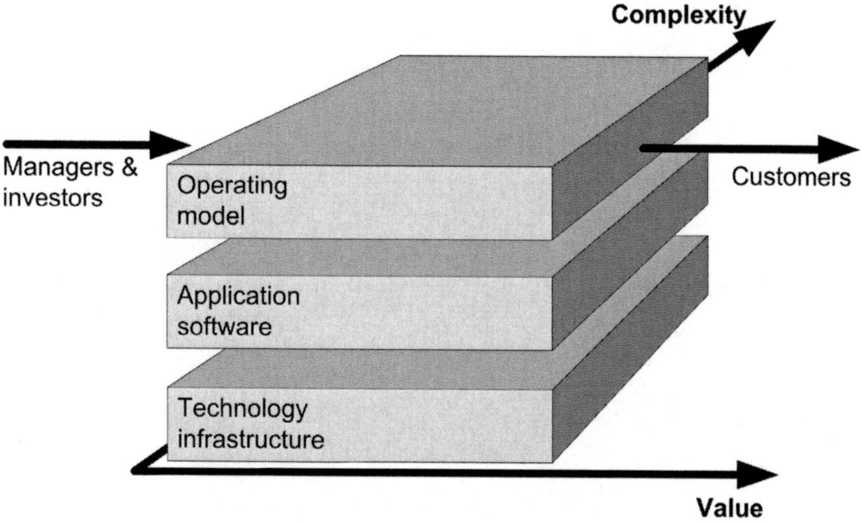

Business strategy is external to the framework, and is not covered by it. The managers of a business are external to its operating model, rather than bound up in it. Part of their job is to review and improve the model, not simply to keep it running, and this is really what strategy is all about: a management team must look carefully at the operating model and adjust it in response to competitive pressures. They must find ways to serve their customers better, or find new customers to serve, or build innovative new products, or any of a myriad of other strategies. Over time, a management team will try to reconfigure the operating model so that it better fits their firm's competitive positioning. The decision as to what changes to make to the model is the domain of business strategy, but this book is about IT, so business strategy is certainly external to our framework.

The external environment is also outside the framework, and that includes the markets in which the business operates. The environment can be analysed using the PEST model – i.e. in terms of political, economic, social and technological factors – or with a wide variety of other tools. But again, this book is about IT, so the analysis of the external environment is again outside our framework.

The external influences on the framework are shown in Figure 22, which is very similar to the previous diagram. This shows the sources of pressure for changes to a business's value map.

Figure 22 – External influences on the business

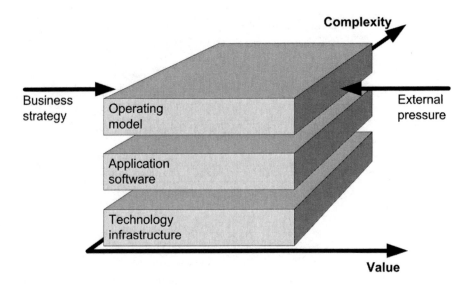

Business can be likened to the scrum in a game of rugby. One set of forwards (the management team) are shoving in the direction they would *like* to take the business; but the actions of their opponents (their competitors in the market) and the effects of external environmental factors may push the business in a completely different direction. The actual changes which occur to the firm's operating model will be partly directed by the management team and partly by circumstances.

Henry Mintzberg showed how this applied to Honda's success in the American market for small motorcycles. Countering the commonly-held belief that this success was all part of a grand strategy devised in Japan, he observed that, in fact, Honda's managers had never intended to sell small motorcycles in

America – they had come to sell the sort of large machines that they (and everyone else) assumed Americans liked. But the larger motorcycles failed to sell, and it was only when the Honda executives themselves were spotted riding smaller 50cc bikes that they started to make some headway with American consumers. So it was an accident that allowed Honda to sell the smaller machines that ended up dominating the US market. Mintzberg said:

> *This is a story of success, not failure, yet [the Honda team] seemed to do everything wrong. True, they were persistent, their managers were devoted to their company, and they were allowed the responsibility to make the important decisions on site. But when it comes to strategic thinking, they hardly appeared to be geniuses.*[47]

Actually, what the Honda managers did do right was to react quickly to external market pressures. Their initial strategy was failing, so they adapted quickly and eventually succeeded. No business operates in a vacuum: a company's operating model evolves over time under continuous competitive pressure, but so too do those of its competitors, in a kind of arms race, and the process is never-ending.

At the time of writing, American Airlines is in the middle of rolling out a wi-fi service that will allow its customers to connect to the internet during their flights. In the short term, this should help the business to compete by offering its customers a good reason to select American over other airlines, in a brutally competitive industry. But American's competitors will not stand still – other airlines are also installing wi-fi systems. In the long term, it is likely that in-flight internet access will simply be a part of travelling by air, that passengers will expect it, and that all airlines will have to offer it. There will be no premium for offering the service, and American Airlines will be back to square one. Competition is trench warfare, and IT's contribution can only ever give a business a temporary advantage.[48]

In this never-ending competitive arms race, managers have a two-fold task: to run the day-to-day business, but also to step back and continuously improve the business operating model. And IT systems have a critical role to play in this process of continuous improvement, because they provide the information that managers need in order to plan the future. This is shown in Figure 23.

Figure 23 – IT helps both to manage the present and to plan the future

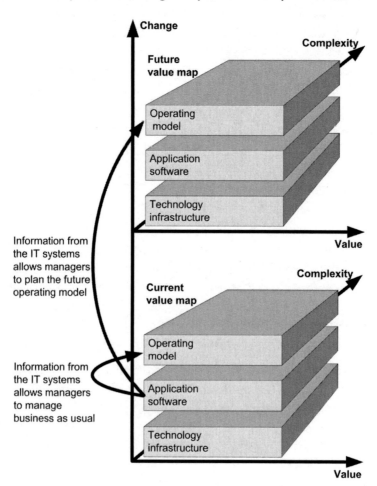

Coca-Cola uses SAP across its worldwide operations, and the ERP system plays a crucial part in continuous improvement. Becky Glenn, SAP Global Integration Manager at Coca-Cola, said: "mySAP (ERP) Financials, along with SAP Business Intelligence, allow us to compare one part of the world to another, one brand to another, to answer crucial questions about our business, and continually improve our competitive position."[49]

The framework is well suited to discussing the evolution of systems over time, because it allows for a conceptual view of the future operating model. In fact it is possible to plot several stages in the evolution of the model, and to consider alternative scenarios. Figure 24 shows how the framework might be used to debate that evolution.

Figure 24 – How the value map evolves over time

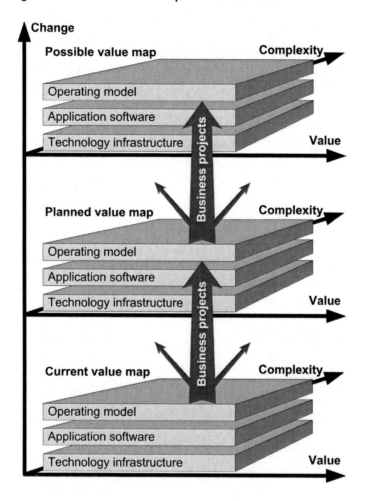

In a sense, the change dimension can be used to show a series of related, but time- and choice-dependent, options. This sort of decision tree can easily be adapted to scenario planning. Looking ahead to Part IV, we can also use scenario planning to think about value. Often, the best way really to understand the value of an IT system is to think about *not* having it. And the schematic approach shown above can allow a firm to set out the different 'alternative futures' in a way that makes these choices clear.

The same process of continuous improvement also drives changes to IT systems. As a consequence, software applications tend to become ever more complex and granular. A small company might have a simple billing process, perhaps creating each invoice individually. But as the company grows, the invoicing process will become more and more complex, in order to deal with multiple currencies, local taxes and so on. Software is complex because it mirrors how processes work and what people do, and they operate in a competitive environment of continual refinement. This is why an apparently simple process like invoicing can require very complex supporting software. Software gets more and more complex because it has to. Under the normal, static view of information systems, the question is often asked whether the systems are 'fit for purpose'. But if a software application is continually becoming more complex in response to new demands, it really makes little sense to ask this question. The software might be fit for purpose now, but it will have to evolve over the long term in parallel with the operating model that it supports. A static view of IT systems is not effective in managing a continuously changing business. Thus it is not sensible to talk about IT systems being fit for purpose.

A good example is provided by Brandix, a garment manufacturer based in Sri Lanka. With customers that included many high-street retailers, Brandix was under pressure to improve turnaround times and to become more flexible in adapting quickly to late changes in customer demands. In an interview with *ZDNet Asia*, Brandix's executive vice-president, Lalith Withana, said:

> *The internal ERP system we had before was a set of piecemeal modules, all very disjoined from one another. Customers wanted connectivity to our [manufacturing] status, and we had to keep up with demands like overnight changes on the cutting floor. The garment manufacturing industry is seeing turnaround times shrink from over 30 days to a single day. Where the customer is demanding fast turnarounds, your planning modules have to change and be much smarter. A small mistake can be very costly because we produce in high volumes, so your [IT system] has to be flexible to accommodate changes and manage the process.*

The installation of a Lawson ERP system was intended to help address both these issues, and allow Brandix to react to new customer pressures. Withana went on to say:

> *In the fashion industry, you can't compete just on price and quality any more – that is standard. Keeping up with the competition requires a system. You don't realize this till it hits you, that you need a good system whether you like it or not.*[50]

Brandix's information systems might have been 'fit for purpose' previously, but they needed upgrading and improving because the market had moved on.

Continuous changes to a firm's operating model provide the context for the portfolio of IT projects – the cornerstone of IT strategy. But the framework suggests that the strategy is only loosely aligned with business strategy, because much of it is concerned with building future flexibility. This idea, that IT strategy might not be aligned with business strategy, is practically a heresy in today's world. According to Paul Strassmann, "Aligning information systems to corporate goals has emerged as the number one concern over the last five years in surveys of information systems executives."[51] And a survey in 2008 found that: "CEOs, CIOs, and top enterprise managers continue to cite IT-business alignment as the number-one hurdle they face."[52]

Perhaps the dot-com boom is part of the reason for a certain degree of confusion between information technology and business strategy. From the late 1990s up to 2001 entrepreneurs hunted for ways to build new businesses or reconfigure old ones for the internet. With the benefit of hindsight, we know that most of these new business models were flawed: for many businesses things went on pretty much as before, with the web simply adding another sales channel. But the idea still lingers that IT has an intimate connection with business strategy. So the point made above, that IT and business strategy are not necessarily closely aligned, may need some justification.

As a quick test, we can review the annual reports of listed companies. A company's annual report sets out its historical financial results, but also its future plans and prospects. And it is very striking how little mention there is in these reports about the IT requirements of the business plans. As we have already seen, this is partly because such plans are developed at the top level within the company, while technology-enabled changes are executed at the front line of processes and people, and it is only at this level that you can see the interactions between the business's needs and the technology platform. For example, a bank might have a broad strategy which says, 'We must get closer to our customers'. This is hard to translate into technology terms. But suppose the bank decides that, to help in implementing this strategy, it will change its operating model by appointing two additional customer-services staff in each of its retail branches. Suddenly there is a target that the technology team can aim at: each branch will need extra PCs, connections to the corporate network, application software, training and so on. The IT team can now make plans because the business strategy has been operationalised.

The discount retail sector in the USA is a tough market, with competitors that include Wal-Mart, the largest retailer in the world. But in that market, Family Dollar Stores has managed to survive and succeed. Part of its success has been derived from a close partnership between business and IT teams, built on a jointly-developed vision of their 'Store of the Future' and on the technology framework required to deliver on key strategic goals. In an interview with *CIO* magazine, Family Dollar's CIO Josh Jewett said: "We birthed the store of the future project because we didn't believe our existing technologies were necessarily going to drive those goals." The strategic goals were set out, and then the technology framework was designed to deliver them.[53]

A second reason for a lack of alignment between business and IT strategy is that much of a firm's strategy may not be concerned with changes to the operating model at all: in fact, many strategies require no such changes. A good example is the sort of pricing war that retailers periodically declare on one another. This might be a key element of their business strategy, but there are no significant implications for the operating model, since pricing changes can easily be accommodated within the current model. Other strategies of this nature include new-product launches, changes in product mix, advertising campaigns and changes in key suppliers. None of these require significant changes to the operating model.

But the framework also offers a key insight that cuts to the core of the whole conundrum of investing in IT. Many business projects, perhaps even the majority, depend on IT systems *that have already been implemented*. Businesses are exploiting capabilities that have already been created by previous investments in information technology, as shown in Figure 25.

Figure 25 – Most business projects exploit existing IT capabilities

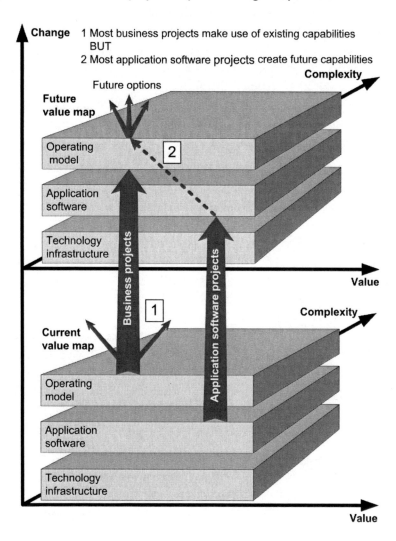

This is perhaps the most important insight that the framework offers. A lack of alignment between business strategy and IT strategy may be caused by nothing more than a difference in *timing*. Much of business strategy makes use of existing IT capabilities, but much of IT strategy is concerned with building future capabilities. Tomorrow's business projects may depend on today's IT investments, but today's business projects actually rely on *yesterday's* IT investments. The two are out of step.

As a crude generalisation, management teams tend not to allow for this delay in the effects of IT investment. For example, a new ERP system will have a likely life span of several years, possibly even decades. It will enable changes to people and processes over the whole of its lifetime. But when these innovations are introduced, nobody will assign any 'credit' to the original investment in the ERP system – the project will be long since finished and forgotten. An accountant would think it a joke if it was suggested that part of the value of these future innovations should be attributed to the original investment in the software application.

But perhaps there *is* a way to handle this. What if firms at least *acknowledged* the contribution that IT has made, through the project-approval process? How might this work? Generally, when managers set out the business case for a new project, they will also analyse, as part of the process, the resources required for it. Sometimes a project will need explicit IT support, so this will be an element in the resource requirements. However, suppose that the IT platform is *already* flexible enough to support the business project. In that case, it goes unmentioned. But perhaps it should be noted: perhaps, every time a business does something that is enabled by a previous IT investment, it is recorded. The previous decision to invest in a flexible IT application would at least be acknowledged and fed back into future decisions, which would help managers understand IT's role as a creator of future flexibility. They would see that yesterday's investments in IT systems have allowed for today's business innovations. A simple tick in a box could go a long way towards helping a business understand the real role of IT.

In summary, business and IT strategies are not necessarily closely linked. Personally, I would go further than this. I am not even sure I know what an IT strategy actually *is*. The role of IT *systems* is to support a company's operating model, and the role of IT *projects* is to support and drive changes to that operating model. Instead of talking about IT strategy, companies should focus on the key applications which will support their current operating model and will allow it to be adapted to future requirements. This focus on the key applications should then drive other aspects of the IT function. I would replace IT strategy with 'application plans' and leave it at that.

Part III
Applying the Framework to IT Projects

Part II introduced a new framework for thinking about IT investments. The first key feature of the framework is that it shows the layered nature of the interaction between information systems and business. Computer systems can support a business but they do not *directly* add value.

The framework deals explicitly with organisational structure by depicting each layer in two dimensions, representing value and complexity. It offers insights about the benefits of integration projects, for example, that can only be understood when viewed through the lens of organisational structure. The two-dimensional value map offers a holistic perspective on the IT project portfolio, since it gives a perspective on IT projects that is fundamentally based on a high-level view of the business operating model.

Part II also discussed the dual nature of IT in both supporting the current operating model and helping to build the future one. The framework shows the dynamic nature of investments in information systems, many of which will enable future and unknown options. It was suggested that there is no meaning to the phrase 'fit for purpose', because information systems have to change as the business changes under constant competitive pressure. And a clue was given that the development programmes for the three layers (business, applications and infrastructure) may be running to three different timescales, and may thus be only loosely coupled.

Perhaps the most important insight offered by the framework is that many changes to the operating model involve the exploitation of previously-created technology capabilities. If alignment between the business and IT functions is elusive, managers should focus on their key applications instead of trying to manufacture an IT strategy artificially.

What has yet to be shown is how the framework might work in practice. (Indeed, it has yet to be shown that the framework is useful at all!) In Part III we will look at some of the typical IT projects about which businesses must make decisions, and how the framework can help. This part of the book is organised around four generic types of IT project; these will probably sound unfamiliar, since they are not adopted in other approaches to IT investment. In the process of discussing these categories of IT project, we will draw out a number of key decision factors for each.

Chapter 7
Improving the Way a Business Works

In the first category of IT projects are those which are aimed at improving a business. They enable new automation and information benefits, and allow a company's people and processes to work better. The key point about such projects is that the company's operating model is being *qualitatively improved*. Such projects improve the way the business does business. This might sound obvious, but it does not apply to all IT projects, and we will come to some of these shortly.

Bespoke Software Projects

Let's start by looking at bespoke software developments. This type of project has one interesting feature for our purposes, in that the design of the software should reflect the needs of the business. That is what bespoke software is: the aim of writing a bespoke system is to create software which is customised to a company's requirements. It is how computing used to be done in the past: there were no packaged applications, so companies built their own software solutions from scratch, and each one was different. Here is one case where business plans and IT plans *are* closely aligned.

For example, Deutsche Bank implemented a bespoke system for foreign-exchange (FX) trading called autobahnFX. The system is based on a bespoke Java server, and is now an integral part of Deutsche Bank's offerings in FX. It is an example of how a bespoke application software project was necessary to support changes in the business's operating model (supporting clients in conducting FX trades). The benefits are not easily measurable, but they can at least be identified. In the case of Deutsche Bank, the autobahnFX platform provided a new and much richer flow of information to customers.[54]

Although the primary aim of a bespoke software project is to enable immediate changes to the operating model, such a project may also build capabilities for future options. Software engineers often have to balance the need for building a new system quickly against the need for maximum flexibility in the future. At one extreme, they can write software which is a 'quick fix' but which will not be extensible; at the other, they can write software which acts as a flexible development framework for the future, but which takes much longer to write. As a very broad generalisation, the need to develop software quickly usually wins. Future flexibility is a secondary goal, and the balance between current and future capabilities is generally tipped in favour of the current. That said, a bespoke system can often be built and improved upon over a number of years.

Kaiser Permanente is the largest non-profit health-care plan in the United States. One of its key systems is KP HealthConnect, an electronic medical-records database, which was originally implemented to maintain a detailed history of each patient. But once doctors started using it, they quickly realised that the data could be consolidated in order to analyse outcomes to diseases. The benefits of KP HealthConnect can be split between the original planned changes to the operating model, whereby doctors would be able to see each patient's medical history, and the optional future benefits of consolidated data on diseases.[55] Kaiser Permanente's bespoke software project can be mapped onto our framework, as shown in Figure 26.

Figure 26 – Kaiser Permanente's bespoke software project

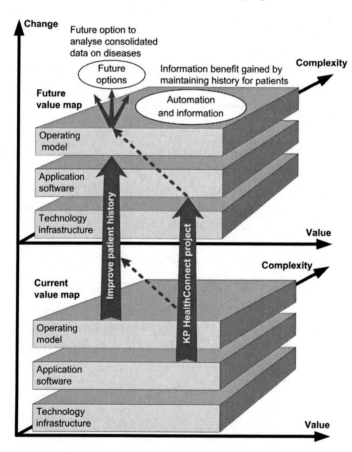

Sometimes a bespoke software application is not optional, but a necessity. That happens when a company's operating model is unique, so that a packaged solution simply cannot work. These cases are rare, and it is normally best to make use of standard packaged software wherever possible. But sometimes a company comes up with a genuinely new way of doing things, and that will require a bespoke application. After all, if a business adopts a standard package that forces it to operate like its competitors, how can it gain any competitive advantage? This point was well made by the former CEO of Merrill Lynch, Stan O'Neal, when he said: "Build whatever can differentiate you from your competitors, but buy the rest."[56]

For example, Chubb believes that it has gained an advantage in the highly competitive insurance industry by building a bespoke system for online collaboration with its agents. Chubb is not using its system simply to drive costs down and support its business processes, but explicitly to gain a competitive advantage.[57]

Businesses in such sectors as online retailing, banking and telecommunications rely particularly heavily on information technology in building a competitive position in their markets. Such businesses should expect to invest in a higher proportion of this type of project, since it is bespoke systems that are the most likely to sharpen their competitive edge. Their IT project portfolios will also be more closely aligned with their business strategies, if only because they will include a higher proportion of bespoke projects.

One final aspect of bespoke systems deserves to be mentioned. A bespoke system can evolve over time in parallel with changes to a company's people and processes. By contrast, a standard system frequently constrains the operating model to fit with the packaged software. Although bespoke software projects are conventionally seen as risky and expensive, our framework suggests that they can also allow a company to respond better to changing conditions. It allows the software-development cycle to be shortened, with a series of frequent minor improvements rather than infrequent major upgrades, and this allows for faster turnaround of new requirements, and therefore quicker responses to changing market conditions.

In summary, the focus of a bespoke project is usually on current benefits. It is designed to be a good fit for the current requirements of the business, giving benefits in automation and information. It might offer some future options, but these are usually not the prime focus of the project.

Packaged Application Software

Staying within our general category of improvement projects, we next look at a packaged software application. As with a bespoke system, the objective of a packaged software project is to *improve* the operating model through automation and information benefits. For example, eCourier is a parcel-delivery company, based in London, which implemented a real-time business-intelligence system that gave it up-to-the-minute information on its customers' ordering patterns. The company uses this standard package to give it an information benefit that translates into a competitive advantage.[58]

To take another example, John Deere is a world leader in construction and farm equipment. In 2007, the company decided to install a system that allowed for impact analysis of variable demand across the five manufacturing plants within its Consumer and Commercial Equipment division. This system will help the company to respond more flexibly to variations in demand for its products. John Deere is targeting a key operational lever, flexibility, and building IT systems that support it.[59]

The kind of benefits that eCourier and John Deere gained from their projects were really focused on better information, leading to better decisions. As an aside, information benefits tend to be broader in scope than automation benefits. This makes it harder to estimate their value, and we will look at this in more detail in Part IV.

Benefits can also come from improved automation. Southwest Airlines is a successful low-cost carrier with revenues of $11 billion. Consistently profitable, Southwest has also been given high marks for customer service.[60] The company has not always treated IT as a priority, for example in choosing not to automate its aircraft maintenance. But that has steadily changed as Southwest has been "getting more comfortable with technology", according to Don Harris, director of airport solutions, in an interview with *InformationWeek* magazine.[61] Nowadays, Southwest's aircraft maintenance is underpinned by information systems that include an enhanced optimisation scheduler and tablet computers for its inspectors. Up until 2003, Southwest offered only manually-printed boarding passes to its customers, but then it introduced electronic tickets and self-service kiosks. These measures cut costs, and also made the check-in process quicker for passengers. Better automation has improved Southwest's operating model.[62]

By comparison with bespoke software projects, the benefits are split more evenly between current benefits and future options. On the one hand, a packaged application will usually provide a great deal of new functionality at relatively lower cost, because the software vendor can spread the development costs across all the companies that buy the package. On the other hand, much of the functionality might not be useful in the short term. When buying a packaged software solution, a business will usually have to compromise on its shopping list of requirements – the software might not do everything the business requires in one area, but may be over-specified in another. The organisation might not want these features immediately, but they are there anyway, and they can be switched on in the future. Even a simple software package like Microsoft Excel has many functions which might never be used, but which still allow options for the future.

Chase-Pitkin Home & Garden is a chain of hardware and garden shops based in New York state. In order to keep track of sales information, it implemented an Essbase data-warehouse solution and began to collect data. This led to immediate benefits, but also gave the company options for the future. In fact, it was not until seven years later that the company exploited one of these options, by implementing an analysis tool called SPSS's Showcase Analyzer to dig into the data warehouse. The information that was gained allowed Chase-Pitkin to address a specific performance issue, that of stock shrinkage. Managers could identify where shrinkage was occurring and on which product lines, generating $200,000 in savings in the first year.[63] In Figure 27 we can see a representation of Chase-Pitkin's project.

Figure 27 – Chase-Pitkin's application-software project

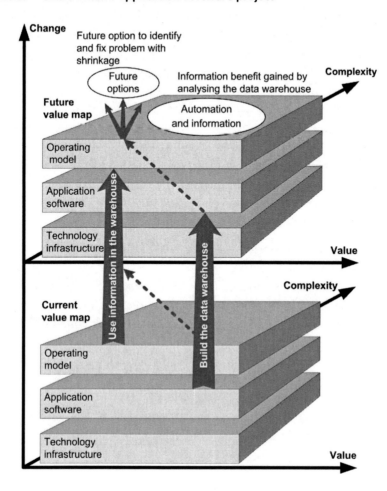

For packaged software, there is a relatively loose fit between business requirements and application functionality, and this affects the software-procurement process. The traditional requirements-gathering approach is fine for bespoke software projects, but works less well for packages, simply because of this looser fit between requirements and functionality. Many of the options that a package enables will be predictable, and can be considered as part of the decision-making process, but many will not even be known. And when people start using the software, the range of possibilities becomes even wider. Many, perhaps most, of the options are not predictable when the software is installed. The traditional process of requirements-gathering thus needs to be broadened to encompass a much stronger focus on future flexibility.

When Cisco Systems upgraded its Oracle ERP system, its two objectives were to increase company productivity and to enable future flexibility. David Murray, director for Oracle Products release management, said: "We knew that an upgrade to Oracle 11i would be a massive undertaking, but we also knew we had to build out the foundation to enable future capabilities."[64] Notice Cisco's emphasis both on the immediate benefits given by improved productivity *and* on enabling future capabilities. Cisco is a company that clearly understands the importance of future options, and the central role that IT plays in enabling them.

To summarise, a packaged system will allow for some immediate changes to the company's value map, but the focus is usually on future options, including those not yet known.

Technology-driven Initiatives

Change is not always driven by the business from the top down. Sometimes the world of technology throws out something that offers new opportunities.

Oil companies spend billions in their search for new oilfields, and much of that money is spent on sophisticated IT systems. One of the largest oil companies is BP, and one of its largest oil fields is Thunder Horse, in the Gulf of Mexico. BP and its competitors use seismic surveys to search for new oil under the seabed. In order to find these new reservoirs, BP must number-crunch the enormous volumes of seismic data that map the underground contours. This requires the sort of computing power which can only be provided by the latest and most powerful super-computers. In 2003, BP upgraded its computers to the latest Intel Itanium processor, which allowed for a quantum leap in computing power, cutting the time taken to survey the underground reservoirs at Thunder Horse to 25 days. Previously, the survey would have taken two years – nearly 30 times as long – a time lapse that had made this type of survey effectively useless to BP's management. The reduced timescale allowed the Thunder Horse project team to repeat their analysis over and over again, gradually improving their picture of the underground reservoir structure.[65] In other words, BP made investments at the level of the technology infrastructure which have *enabled* new business processes that were simply impossible before. This is in keeping with BP's attitude to IT, summed up by then-chairman John Browne in an interview with *CIO* magazine, where he said he viewed IT "not just as a service function ... but as an activity which could change the nature of the business itself."[66]

How can we analyse this within our new framework? The answer is that an innovation at the technology-infrastructure level (a new processor) enabled a new kind of software application (the seismic survey), which in turn allowed for a change at the business level (BP was enabled to use the seismic survey data within its planning timescale). This is shown in Figure 28.

Figure 28 – BP's technology-driven project

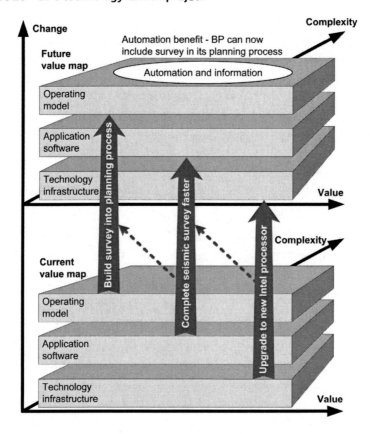

Meanwhile, Cathay Pacific, Hong Kong's airline, has found new ways to connect with its customers through technology. Passengers in Hong Kong can now check in using their mobile telephone, via a wireless internet connection. When they confirm their details, they are given a barcode which downloads to their mobile phone. They can then pick up a boarding pass at the airport by scanning the barcode at self-service kiosks. The whole process is designed to save time and bring Cathay Pacific closer to its customers, and the innovation depends on new technologies.[67]

Innovations can also extend the 'reach' of the IT platform. A good example of this is the Blackberry handheld device, from the Canadian firm Research in Motion. When the Blackberry first appeared, it brought a software application, email, to a part of companies' operating models that had previously lacked any supporting software: people who were working away from the office went without one of the basic communication tools of modern business. The Blackberry filled that gap, allowing remote workers to stay connected to their corporate email networks. This is shown in Figure 29.

Figure 29 – Blackberry email extends 'reach'

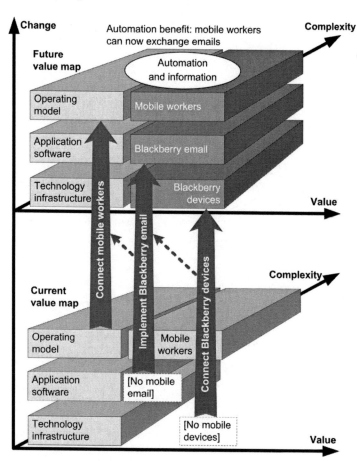

Although the impetus comes from below for technology innovations, the benefits are the same as when a project is driven from above. That is, the business will make certain changes to its operating model, these will be supported by the application-software project and in turn by the technology-infrastructure project, and new options will be created in the future. At a fundamental level, the business operating model is being qualitatively improved.

Chapter 8
Re-shaping the Business Landscape

We have looked at projects designed to improve the way a business does business, but not all IT projects are of this kind. A second type is aimed at managing complexity and business scale. For example, when a group decides that it wants to use a common accounting system across all its operating businesses, the main motivation is not really to *improve* the accounting functions across the group – the real aim is usually *to make them the same.*

This crucial point is often overlooked in discussions of IT value. The standard ROI approach does not cope with complexity at all. But our framework does, because the second dimension allows it to reflect these real-world problems. The value map clearly shows the problems that are caused by multiple applications and technology platforms. In particular, it is broad enough to cope with the implications of mergers and acquisitions (M&A) for information systems, and this ability is not offered by ROI or other approaches to computer-system investment.

Companies spend a great deal of time and money trying to integrate their systems. However, they also spend a fair amount of time trying to split them apart, when they sell a division or a business. In a way, shrinking a business is just as hard as expanding it. And since the former is actually a simpler type of IT project, we will start there.

IT Separation

When a group decides to sell a business or division, it creates a painful headache for the IT team. All the systems that had previously been shared across the whole group must now be separated so that the divested business can operate independently. This typically includes the wide-area network, email system, corporate intranet and many other systems. This sort of IT separation project is complex, expensive and time-critical. It is also somewhat dispiriting, since all the hard work that has gone into integrating the systems in the first place must now be undone.

In 1998, the car-manufacturing giants Daimler-Benz AG and Chrysler Corporation merged to form DaimlerChrysler. As part of the merger, the IT functions were to be combined under a single CIO.[68] But there remained large differences between the two parts of the IT organisation: Chrysler had already centralised and integrated its information systems, while Daimler-Benz allowed each business unit to run its own IT platform. Nonetheless, DaimlerChrysler carried on the hard work of integrating the infrastructure, for example by consolidating a single email system across the worldwide group.[69]

In the meantime, however, trouble was brewing. The newly merged group was struggling, with problems arising particularly from a culture clash between German and American managers.[70] Chrysler was eventually sold in 2007, and the experiment of merging these two giant companies was over. This meant a major separation project for the IT team, who had to undo all the integration work that had already been completed. Daimler's annual report for 2007 says:

> *Because IT was integrated worldwide at DaimlerChrysler, the transfer of a majority interest in Chrysler resulted in great changes. IT separation activities were not yet completed at the end of 2007. In some cases, we decided against setting up duplicate IT systems and arranged long-term service agreements between Daimler and Chrysler. In addition, the new company name and the related corporate terminology had to be installed in all IT systems.[71]*

One part of the integrated group had been DaimlerChrysler Financial Services Americas. To separate the IT systems for this part of the group alone was a mammoth project, involving 126 software applications and external connections to 118 third parties. The team at Daimler was able to complete the work for less than $50 million, allowing for a clean break between Daimler and Chrysler.[72] But the key point is this: there was no aim to try and *improve* the way either Daimler or Chrysler did business. Again, nobody wants to try and improve a business during a separation project, because there are enough problems to be found in simply getting the job done as quickly as possible.

We can see DaimlerChrysler's separation project mapped to the framework in Figure 30. Notice that there are no benefits from improved automation or information, nor from future options. Instead, the benefit of the separation project is to enable the businesses to change scale, in this case downwards. We will look at ways to assess the value of such benefits shortly.

Figure 30 – DaimlerChrysler's separation projects

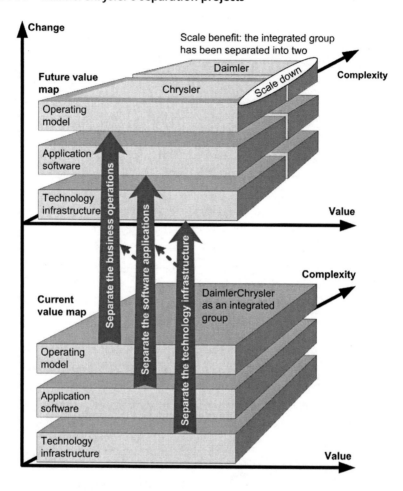

What is happening at the business level can be clearly seen – the operating model is being split in two. That means the applications software and technology infrastructure layers also need to be separated, as fast as possible and with minimal risk. Again, the framework helps with the decision as to whether to approve such a project. There are no long-term options to consider in this case; there are only potential risks and delays. It is impossible to think about the benefits of the IT project in isolation: they will all be derived at the business level, and the IT project is there simply to make the separation happen. To expect an ROI from the IT separation project in isolation is simply unrealistic – yet that is what CIOs are implicitly required to do where they have to demonstrate an ROI for *all* their IT projects. CIOs need to allocate their IT resources across many competing projects, of which the separation project is one, but they cannot do this if the only measurement of value is ROI.

In summary, the benefits of separation projects are almost entirely based on immediate changes in the scale of the operating model. The benefit of the project is to create a viable stand-alone business, and no additional future or unknown options will be generated.

Organic Expansion of a Business

In a way, separation projects are similar to another common type of project, that of organic expansion. Both involve changing the scale of the business rather than making qualitative improvements to it; and neither is concerned with improving it. When a company opens a new branch office or expands into a new territory, nobody wants to be adventurous with the information systems. There are already plenty of risks and unknowns at the business level. As with separation projects, the key is to minimise cost and risk.

Specsavers is a successful chain of opticians, with more than 1,390 stores across the world.[73] It has taken the idea of standard systems to a new level, with a standard deployment model that includes all the IT systems which a new store will need, and allows Specsavers to open a new store in less than 11 weeks. In an interview with *CIO* magazine, Specsavers's CIO Michel Khan said:

> We've just rolled out 100 stores in 100 working days in Australia. We couldn't have done that unless we had a really highly geared process. It's Specsavers-in-a-Box, a complete set of applications covering merchandising, finance and distribution. It's a global framework for expansion.[74]

The corresponding diagram in the model is shown in Figure 31. Again, the benefit of the new-store project is to enable the business to change scale, this time upwards.

Figure 31 – Specsavers's opening of a new store

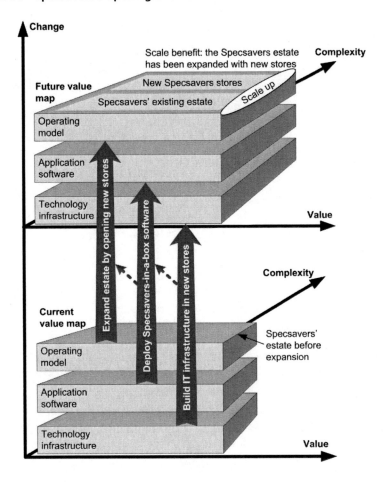

As with separation projects, the changes at the business level can clearly be seen – the operating model is being expanded. The benefits of an expansion project are to be found in the immediate changes to the operating model. The new store will generate new revenues and profits, and that is the underlying motivation for the IT project. Again, it makes little sense to try and justify the IT project in isolation, using ROI. Since a new store cannot operate without IT systems, the benefit of the IT project is the creation of a viable store. No additional future or unknown options will be generated, because the existing systems will simply be extended.

It might be objected that this is no more than stating the obvious: opening a new store is really only a business project that has an IT element to it. While this is true, IT managers are faced with the challenge of balancing scarce resources against multiple needs. They have to balance the opening of a new store against other tasks, with resources and budgets that must cope with all of them. For example, a key software architect might be required to work on multiple projects; how can the CIO decide which one should come first, or where to prioritise his or her efforts, if the only yardstick, ROI, fails to work? We will approach a partial answer to this question in Part IV, but it is useful at least to point out the problem here.

Integrating Applications

Integration is the single biggest challenge for many IT teams. As organisations grow in scale and scope, they inevitably become more complex. Additional complexity comes from M&A activities. When one company buys another, it typically inherits a range of IT systems as part of the deal. The new combined business will have a minimum of two IT platforms, and while there may be benefits to integrating the two, the process takes time, money and management focus. The task of connecting the IT platforms is normally part of a broader post-merger integration (PMI) programme.

Nirvana for a CIO is to have a single IT platform across the whole group. But this is rarely achieved, partly because new acquisitions bring new systems which need to be integrated. An aggressive business that grows continually through acquisition will tend to build up a backlog of integration work. In fact, it is probably fair to say that M&A deals conflict with the long-term efforts of businesses to integrate their IT systems. When Bank of America bought FleetBoston Financial Corporation in 2003, it had yet to integrate its previous acquisition of NationsBank in 1998: customers were struggling to transfer money between the two. And FleetBoston itself had not yet integrated the IT systems after its previous acquisition of BankBoston. The businesses might have become integrated in a legal sense, but the IT systems had fallen far behind.[75]

Whether a lack of integration comes from M&A or simply from organic growth, businesses with such a problem face a tricky question. What are the benefits of integration? Just to repeat, the point of asking this question is to help IT managers decide how to allocate their resources.

Let's start at the business level: when two businesses merge, there is a certain minimal level of IT integration that needs to happen just to get them to work together. This is what DaimlerChrysler did when it integrated its worldwide email systems.[76] This is shown in Figure 32 where, again, the benefits are to be found in changing the business's scale.

Figure 32 – DaimlerChrysler email integration

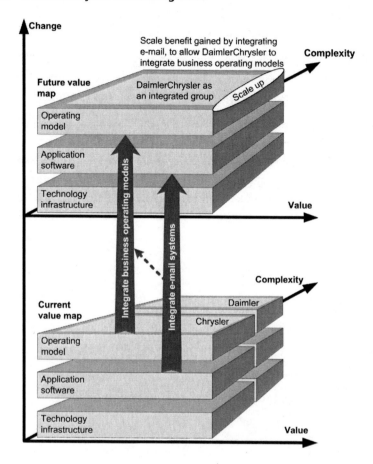

This minimal, shallow level of integration, which can be as simple as email and a few other key applications, is often all that is needed for two businesses to operate as one. But when integration extends beyond the basics of email and websites, and starts to involve ERP and other major systems, it becomes much more complex, because this type of 'deep integration' leads to changes in the way people and processes work.

With each integration project, a decision must be made concerning at least two applications, one from each side of the business. The choices are to pick one or other of the applications as the preferred solution; to implement a new application that replaces both; or to link the applications together using some sort of 'connector software'. This latter approach was adopted by the cosmetic and fragrance business Coty when it integrated its systems with those of

Unilever Cosmetics International. Coty had already achieved the kind of shallow integration described above, but to link the key ERP and supply-chain systems, the IT team used 'middleware' software.[77]

All these approaches involve changes to people and processes, but for the purpose of this discussion it will be assumed that one or other of the existing applications is chosen as the preferred solution for the newly combined group.

In Figure 33 one side of the business already has the 'chosen' application, so the other side of the business must adapt to it – which means changes must be made to the way people and processes operate. But close connections between people, processes and applications software are built up over time, and when the software is changed, these connections must be decoupled and then rebuilt. In simple terms, there is a great deal of disruption and business 'pain'.

Figure 33 – Information and automation links being split then rebuilt

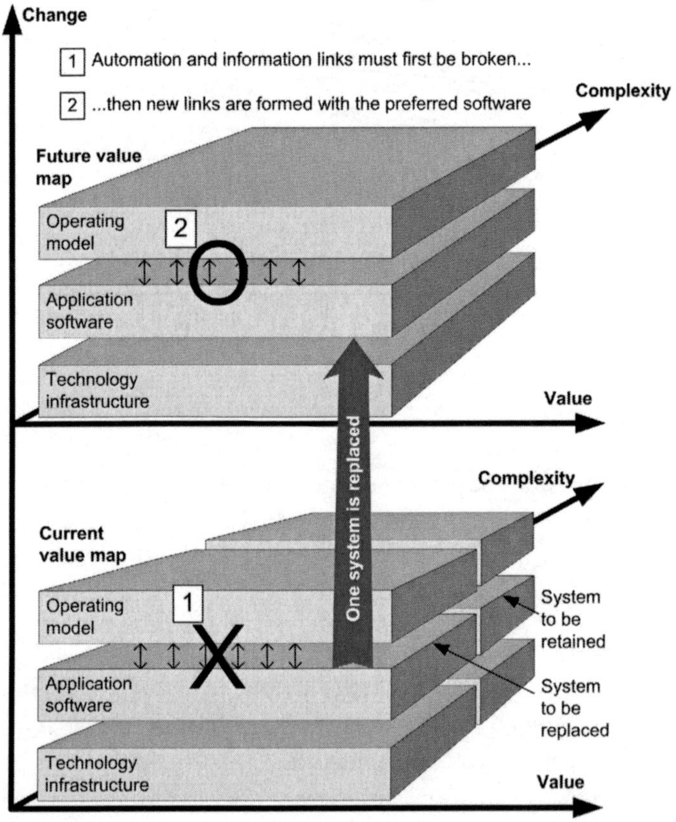

This is a tricky exercise even when the two sides of the group have broadly similar operating models – people still have to learn new systems and processes configured in the software. But the problem becomes acute when the two sides of the group have *different* business operating models. In that case, integrating the IT systems means changing one or both of the models.

Frequently, the management team will state, by diktat, that the individual business units should simply accept the new business processes, on the basis that they represent 'best practice', and that having a single IT platform will cut costs. But according to our decision framework, this is risky: business processes must be changed to suit applications software, rather than the other way round. Instead of concentrating on people and processes first, this approach is an attempt to cut IT costs by imposing a potentially unsuitable IT system on the business.

A business is not of course *obliged* to integrate its applications. This point was well made by Walt Disney's CIO, Bill Patrizio, in an interview with *InformationWeek* magazine:

> *Given our diversity, we are not a homogeneous company. We are in broadcasting, cable, theme parks, Internet, and the cell-phone business. To strive for standardization across the enterprise is not a wise endeavour, because it doesn't acknowledge the unique individual business strategies and processes within the groups.*[78]

Nevertheless, it *might* still make sense to integrate the key software applications (here called 'deep integration'). We may split the decision into two. Recalling why we have software in the first place – to gain benefits from automation and information – there is no reason to assume that the first set of benefits, from automation, will be significantly enhanced through integration. After all, if both businesses are equally productive and efficient, nothing will be achieved through combining their software applications.

What about the benefits that are associated with information? Here, matters are much clearer. When different parts of a group are running different technology platforms, they are less able to share information. At a minimum, that means multiple sets of data to be consolidated and reconciled. But it can also mean differences in the coding of data and in units of measurement, which make it harder to manage the group. So, one benefit of integration is to allow a group to have one source of information, 'one version of the truth'.

The convenience chain Wawa understood this well when it implemented SAP. In an interview with *CIO Insight* magazine, Wawa's CIO, Neil McCarthy, said:

> *We needed one version of the truth. Today, we have a bunch of different data repositories, seven or eight different price books and decision support systems. We want one common repository. Today the store managers see different [data] than marketing and finance. [With the new software] everybody will be looking at the same data.*[79]

Wawa's project is shown in Figure 34.

Figure 34 – Integration at Wawa

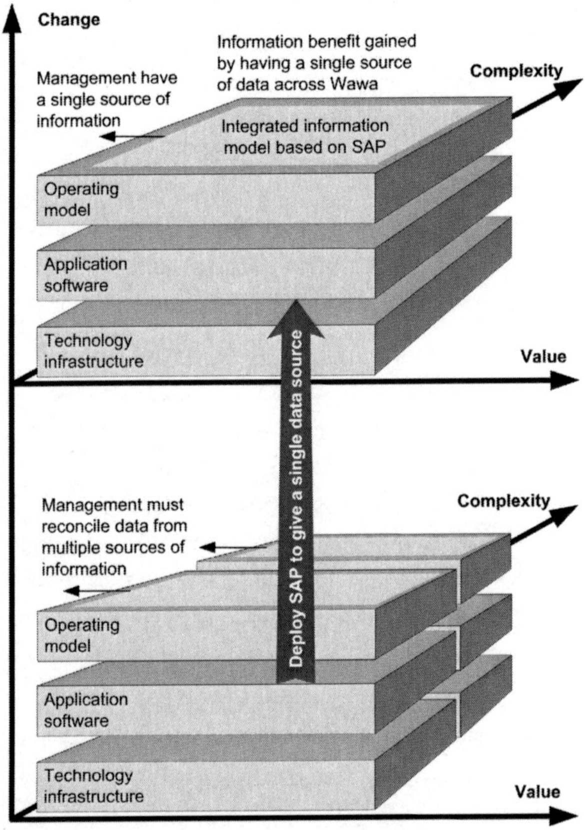

So integration projects do not *necessarily* bring automation benefits, but the improvement in management information will allow the central group team to have a consistent view of the business. This point is not a theoretical one – it is frequently and emphatically made by the managers of local divisions who are being asked to accept a new, centralised IT system, but who know that it will not help them to gain any local efficiencies. The imbalance caused by deep integration is that the pain is felt locally, but the benefits are gained centrally.

A new perspective on integration has been provided in a recent book by Jeanne Ross, Peter Weill and David Robertson. The authors make a useful distinction between *standardisation* and *integration* of business processes.[80] For example, McDonald's has a high degree of process standardisation within its restaurants across the world. In terms of process automation, what works in a restaurant in the USA can usually be replicated in a restaurant anywhere in the world. But customers are not shared between restaurants, so there is no need for unit-level data to be shared. Instead, McDonald's can focus on building standard processes and replicating these across the world without worrying about sharing unit-level data. Business-process standardisation is all about automation.

By contrast, banks offer many different services to their customers, including investments, foreign exchange, lending, mortgages and current accounts. They cannot simply replicate processes across these different business areas, but they do need to share data. One of the reasons is that customers are shared across the business (unlike McDonald's). A customer with a current account might well prefer to buy a mortgage from the same bank, and the bank needs to make that as easy as possible; so it must focus on sharing information rather than on automation. Business-process integration is all about information.

This provides a useful tool for thinking about whether and how to integrate systems. If business processes are standardised, the same automation approach will work across the group. This shifts the decision in favour of common software applications, but not necessarily in favour of common data: the same system might be used in separate local databases (the IT term is 'instances'). But if business processes are integrated, the focus should shift to common data, and not necessarily common software applications. Interfaces and data feeds between disparate systems may be more appropriate.

Integration is also tied up with the fundamental raison d'être of businesses. As long ago as 1937, Ronald Coase asked an interesting question: why do companies exist?[81] Why do we not all become freelance workers, and collaborate on projects as and when required, being paid by the customer for the work we do? After all, people give up a large fraction of their time to work, and accept certain limitations on their freedom (very few people really enjoy turning up for work on a Monday morning).

Coase's answer was that firms only exist because the cost of transactions is reduced within them. In other words, firms exist because the alternative is a complex world where people would have to deal with each other as freelance contractors on every single transaction. The overhead for this complexity would be enormous: every single transaction would involve a search for the right freelance worker and a negotiation on price and contract terms. Coase wrote this long before the computer age began, but it has important implications for information technology, because IT systems dramatically reduce the cost of information, and therefore transactions. It should be easy and cheap for people to work together within a modern company, because everyone is connected together via the IT systems.

Coase also observed that the cost of transactions can set the size of the firm. At some point, a firm will grow large and complex enough for it to become cheaper to carry out transactions through the external market rather than through the internal mechanisms of the firm. It seems likely that this point can be extended by use of information systems, since they provide cohesion to a large enterprise, thereby reducing the internal transaction costs. In that sense, information projects have the effect of *defining* the boundaries of the firm. Having an integrated information system cuts to the heart of what the company actually *is*.

So an applications-integration project can reduce the cost and complexity of transactions within a firm and, in a sense, define the very boundaries of the business. As a rule of thumb, application integration makes sense for any business that must be managed as an entity rather than simply as one business within a portfolio.

More subtly, groups which have failed to integrate their systems find it harder to make automation changes across the business, because they have to configure the changes across multiple systems. For example, when a group wants to modify its supply chain through automation, it needs to make changes to the underlying supply-chain applications software. But if it runs several different such applications across different businesses within the group, these must all be changed, which is a complex task. So by integrating applications a business can generate better future options for continuous improvement.

Volvo Construction Equipment (Volvo CE) is a subsidiary of Sweden's Volvo Group, with sales of SEK 56 million in 2008. Volvo CE needed to ensure that its 16 production plants worked together to maximise output and quality, and to allow the company to manufacture any part at any location. Part of this involved the installation of a common manufacturing system across all the plants. This software allows Volvo CE to configure the detailed processes at plant level. And by having a common system, the management team can measure the effectiveness of methods across plants, allowing for best practice to be shared. Interviewed by *Manufacturing Business Technology* magazine, Scott Park, Volvo CE's CIO and senior VP of processes and systems said: "We need to be able to make any product at any location. We need to be able to measure the effectiveness of methods used at different locations. Best practices can then be shared." In other words, Volvo CE's common manufacturing system gives it a platform for continuous improvement at its manufacturing plants across the entire business.[82] This is shown in Figure 35.

Figure 35 – Integration gives a platform for continuous improvement

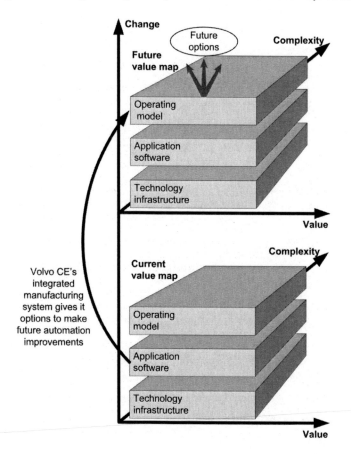

Of course, a business might be able to make significant cost savings by merging its operations. These business-level savings are usually part of the reason for merging two companies in the first place. But the case for applications-integration projects should not be made solely on the basis of *cutting direct IT costs* – the benefits of an integrated information platform are likely to far outweigh such savings. These benefits come from support for basic integration, improved top-level information, and improved options to implement automation programmes. Much the same can be said for restructuring projects, where the organisational landscape changes without being scaled upwards or downwards.

Integrating the IT Infrastructure

What about the IT infrastructure? We have so far skirted round this issue, although it might be noticed that applications integration tends to require an integrated infrastructure. But the approach can be a bit different to that adopted for applications because the technology infrastructure does not touch the business, so there is no need to break and rebuild the complex connections between people, processes and software. That means firms have a much freer hand in integrating the technology infrastructure to reduce costs. Indeed, they can often do this without integrating their software applications, simply by running multiple applications on the same technology infrastructure.

For example, Associated British Foods (ABF) decided to build an integrated IT infrastructure, but to continue running decentralised software applications. Interviewed by *ComputerWeekly* magazine, ABF's IT Director, Trevor Hanna, said: "We are here to give the business units exactly what they want. They are all different, which is why we can get few economies of scale. That's why we have no intention of going to a single ERP system."[83]

In the M&A context, the integration of technology infrastructures is often a key deal driver. When two businesses are brought together, the argument goes, why should they have two sets of technology-infrastructure costs? Would it not be cheaper to have one data centre, one network and one set of licences, with volume discounts? In fact this is a good reason to integrate the technology infrastructure. At this level, matters are much simpler; the only question is whether the benefit of reduced operating-cost savings outweighs the capital cost and risk of the project. This is a familiar problem, and one which most businesses can address. Combining infrastructure is simply a good idea, and the question should be not why but when.

However, there are further benefits to integrating the technology infrastructure, in the options and flexibility that integration offers in deploying applications software. HSBC is a worldwide banking giant, with headquarters in London, revenues of $89 billion, and 296,000 staff spread across 86 countries and territories.[84] HSBC's IT operations are on a similarly vast scale, with annual IT spending of $4.4 billion in 2005. The bank embarked on a major integration exercise as part of the 'One HSBC' programme, and part of this entailed simplifying and rationalising the IT infrastructure. This in turn involved a reduction in the number of data centres and in the range of applications, and a focus on key territories, such as the United Kingdom for hosting and India for software development.[85]

Although the integration programme was partly designed to reduce the bank's operating costs, this was not the primary goal. The technology integration programme has also given HSBC options to deploy software applications into new territories much more quickly and cost-effectively than would be the case with a disjointed infrastructure, thereby increasing reach and flexibility. For example, the roll-out of new credit cards in Indonesia was handled from London at relatively low cost. Presumably HSBC could have launched the new cards without an integrated technology infrastructure, but it was able to do it faster and at lower cost, so this business option was certainly made easier. Ken Harvey, HSBC's Group Chief Technology and Services Officer, said: "That's the real value of having one network and running it as one global network – incredible leverage."[86] The key point is that by providing increased reach and options, the integration of HSBC's technology infrastructure achieved much more than simply a reduction in direct IT costs. It seems doubtful that these benefits would have shown up in an ROI analysis. Taking all this into account, the benefits from integration can be presented as shown below in Figure 36. The business case for technology-infrastructure integration projects can be expressed in terms of reducing the cost of operating the IT platform, and in those of expanding the range and reach of the applications software that can be deployed.

Figure 36 – Integrating the technology infrastructure

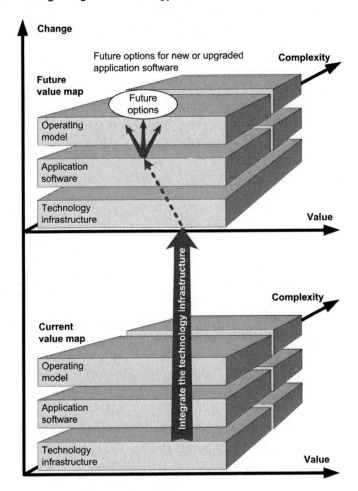

Chapter 9
Big Bang: ERP and Other Mega-projects

A third major type of IT project is concerned with integrating a firm's value chain.

ERP was born in the 1970s, when SAP AG created the first software system designed to work across all of a business's functional areas: a single system provides a single source of data across the business. This was fundamental to SAP's vision, as set out in a paper by Timo Leimbach:

> *The aim of integrated software was to use a common, logical database for all the various applications. After entering an order, for example, users could then automatically initiate the corresponding material requirements planning, billing, and entry into the company's accounting thanks to the ability to generate the data from a single source.*[87]

ERP systems can significantly improve a company's operational efficiency. For example, Unilever's chairman, Patrick Cescau, credited the implementation of SAP's ERP system with contributing to a growth in sales of 7.4 per cent. This was achieved through better allocation of resources, faster decision-making, and lower costs.[88]

But ERP projects are also huge, risky and expensive. In 2002, the discount retailer Kmart Corporation was brought to the brink of bankruptcy by problems with its ERP system, part of the problem being a $195 million write-off related to supply-chain systems.[89] And Hershey lost $150 million in sales when its implementation of new enterprise systems went wrong and it was unable to fulfil demand during the critical Halloween period in the USA.[90] But Hershey recovered, implementing an upgrade to mySAP.com in 2002, on time and to budget.[91] However, some companies never recover: when FoxMeyer Drug went bankrupt, it blamed a failed ERP implementation.[92]

Because of the complexity involved in such large companies, it is easier to see how ERP works in a small business. Artisan Hardwood Floors, a small company with 37 staff based in Texas, decided to install SAP in 2007. The objective was simple: to integrate the different parts of the business via a common information system, so that the flooring-installation business, timber yard and distribution arm could share a single database. That allowed the management team to see a much clearer picture of Artisan's financial and operational situation.[93]

The principle behind ERP systems is shown in Figure 37.

Figure 37 – How ERP projects add value

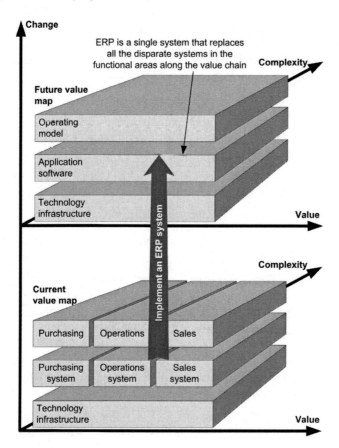

By implementing one system along the whole value chain, a company can identify and fix problems much more easily. When Peak Scientific Instruments installed an ERP system it was able to reduce the number of finished items in store from 40 or 50 to one or two. Interviewed by *Computer Weekly* magazine, the operations manager, Ben Cotton, said: "That was an immediate saving of £400,000 that was no longer tied up in inventory. [The ERP system] ensures we are driven by orders, pulling product through the system."[94]

ERP programmes can offer great benefits, but are also risky. That is because such implementations are fundamentally two programmes in one. We have talked about integrating the functional areas that make up a company's value chain. But a large ERP programme usually involves integration of different businesses *across* a group (if only because any big company is likely to have more than one existing core application). So it is effectively an integration project too, with all the attendant benefits and risks. These dual objectives of ERP programmes are shown in Figure 38. This perspective may help to explain why ERP projects are so risky, because a business which embarks on a large ERP implementation must cope with the changes required to achieve both objectives at once.

Figure 38 – ERP projects usually aim for two kinds of benefit

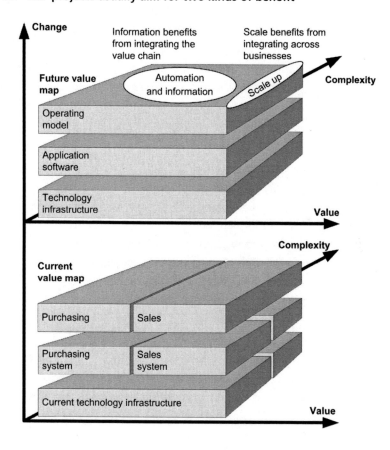

It may seem surprising that any large ERP programmes ever succeed. They always involve a measure of grit and determination. Based in Switzerland, Nestlé SA is a giant nutrition, health and wellness group, with sales of CHF 110 billion, and 283,000 employees all over the world.[95] In 2000, Nestlé decided to implement SAP, at a planned cost of $280 million,[96] across 80 per cent of Nestlé's worldwide operations by the end of 2003, as part of a broader initiative called GLOBE ('Global Business Excellence'), at a planned cost of CHF 3 billion.

What were the objectives of Nestlé's SAP and GLOBE programmes? Chris Johnson, the manager responsible, spoke at an investor presentation in 2005: "The first objective [is] to harmonise best practices across the Nestlé group, the second to establish data standards and data management and the third to standardise the systems." Before the GLOBE programme, a Kit Kat Chunky chocolate bar might have had ten different product codes and descriptions in ten different countries. Johnson went on:

> The first two [objectives] are the most important, the first two are those that really enable the benefits. The last one, the standardised systems, [is] important because [those systems] support the best practices, they support the data and they also enforce them.

So Nestlé used the GLOBE programme both to integrate its systems and as a means to improve its operating model, by standardising data and best practices. The programme took longer than planned, and the original objective, of implementing SAP in 80 per cent of Nestlé's operations, was re-scheduled for 2005, two years later than planned. The biggest hitch was in South Africa, where the SAP roll-out took place against the background of a labour strike. Other problems occurred with high-level decision-support reports, where some 200 reports were not ready when SAP went live. Johnson spoke about these problems:

> And by decision support I mean at a higher level, let's say at management level having consolidated and aggregated reports, strategic and analytical reports to help run the business. And the reason why we're struggling a bit with this [is] because our focus has been really to make sure that all the transaction stuff is working well ... the information at the transaction level; in other words, if you're working on a factory floor the information you get is great.

But despite these teething problems, Nestlé stuck with the SAP programme and the broader GLOBE initiative, and by 2005 benefits were being felt across the group. In Russia, the number of raw materials was reduced; in South East Asia, the year-end closing process was accelerated; in Israel, market share was increased; and in France, new business opportunities were found by sharing information on customers across divisions. But more than that, Nestlé's GLOBE initiative has allowed it to operate as a global entity, to implement other programmes that would not have been possible without integrated systems. As Johnson said:

> *I mean we talk about global businesses like nutrition. I think it would be very, very difficult without GLOBE to really leverage nutrition as a global business in the future, taking advantage of what's specific in the market and leverag[ing] [that] with [the] global benefits of nutrition.*

And GLOBE has enabled Nestlé to do things that were not part of the original plan at all, enabling options that were not (and perhaps could not have been) predicted when the decision to go ahead was taken. Chris Johnson again:

> *And we're continually finding areas that the markets are learning and discovering about. … [T]he focus is starting to tip already to how can we leverage GLOBE, not just put it in. Up until now it's been OK just to put it in … that's been my job. But I think there's some exciting stuff actually, if you can think creatively about what could be leveraged or enabled by GLOBE.*[97]

Similar challenges were found at Celanese, a $7-billion chemicals business, which decided to integrate its multiple SAP systems into one instance of mySAP. In effect, the mySAP programme became an exercise in integration. In an interview, Celanese's Integration Manager, Russ Bockstedt, said: "The real project is, we're trying to become one company, and this is how we're doing it." Celanese was using the mySAP programme to drive integration in what had been a traditional holding-company culture.[98] But the mySAP integration programme also had other objectives: in Celanese's annual report for 2004 the company said that it "began implementation of a company-wide SAP platform to reduce administrative costs by eliminating complexity in information systems and to provide for ongoing improvement in business processes and service."[99] The integrated SAP system was to act as a platform for ongoing improvements, as well as a means to cut direct IT costs.

We can summarise the business case for ERP programmes as follows. They provide benefits in three areas: they enable immediate changes to the operating model; they create options for future change; and they offer benefits through integration. By integrating systems, the business is much better placed to create and exploit future capabilities, because there is a common platform from which to operate. An ERP system creates a large variety of future options (often unanticipated at the outset) for improved automation and information.

Chapter 10
Building and Maintaining the Platform

In the previous three chapters we examined projects that are at least partly aimed at some sort of immediate business benefit, whether derived from improving the business operating model or from scaling it. In this final section, we will look at projects concerned with building a platform for the future, while reducing business risk. We will also cover the operating spend of maintaining business-as-usual.

Technical Upgrades of Application Software

We look first at 'technical upgrades', where the company's primary objective is to reduce business risk. The objectives of such upgrades are to bring the software applications up to date and to make them more supportable. Notice that the risks in this case are not to the *project*, but to the business itself.

For example, a manufacturing plant that uses an old automation system is exposed to a business risk that the system might fail. The company might decide to perform a technical upgrade that reduces this risk, even though it will gain no immediate *functional* benefits from the upgrade. Actually, bearing in mind that the company's primary rationale is to reduce business risk, it may not want to make *any* immediate changes to its business operating model. However, a technical upgrade will also give a business more options in the future, just as any other package of software applications will. The company gains these additional functions whether it wants them or not.

It was to reduce risk that the insurance company Standard Life decided to replace the software that calculated commissions for its salespeople. The systems had been developed in the 1970s and were reaching the point where they were hard to support, representing a business risk to the company. By replacing the old systems with new ones, Standard Life was able to mitigate the risk, and at the same time to become compliant with new regulatory requirements.[100] In other words, the new systems gave Standard Life new capabilities at the business level. This is shown in Figure 39.

Figure 39 – Standard Life's technical upgrade

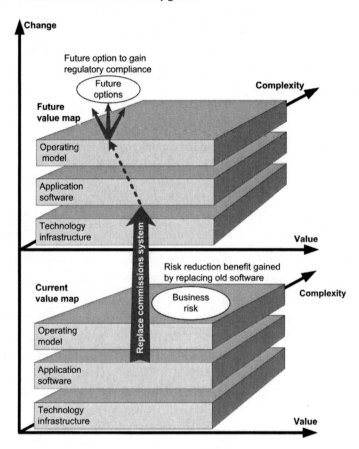

In the extreme case of 'legacy' systems, where the software has not been upgraded for a long period of time, an upgrade project might be needed because the system is blocking changes to the operating model – it is reducing the range of options that the business has for making changes.

At the time of writing, Union Pacific Railroad is upgrading its old mainframe system. The company adopted IBM's mainframes early on, when they first appeared in the 1960s. It has adapted the technology as far as it can, but as business requirements have changed this has become harder and harder, and in the end Union Pacific decided to replace its mainframe with a more modern system. It plans to complete the job by 2014. The mainframe is a legacy system that was blocking many of the business-level changes that Union Pacific would like to have made, and by upgrading to a more modern platform the company will increase flexibility and generate new options for change.[101]

A technical upgrade will give no immediate benefit other than reducing business risk, but will usually offer options and new capabilities in the future. Because of this, the traditional ROI approach is not very useful; it cannot really value risk reduction, and cannot cope with future options.

Broad Technology Infrastructure Projects

Similar benefits are obtained at the technology infrastructure level. An example of a broad technology infrastructure project might be a company-wide upgrade to the latest version of Microsoft Windows. Such projects are normally assessed in terms of supportability and long-term risk. But as we have seen, technology upgrades should also be seen in terms of enhancing the *flexibility* of each layer of the technology platform, in order to increase the options for the layer above. Many infrastructure projects do not really give immediate advantages: buying infrastructure usually means buying capabilities for the future.

Another factor is business risk, which some projects are solely concerned with reducing. These include disaster-recovery provisions, or most security projects, where the only benefits are in business-risk reduction. The combination of reducing business risk while creating flexibility for options in applications software is shown in Figure 40.

Figure 40 – Technology infrastructure projects

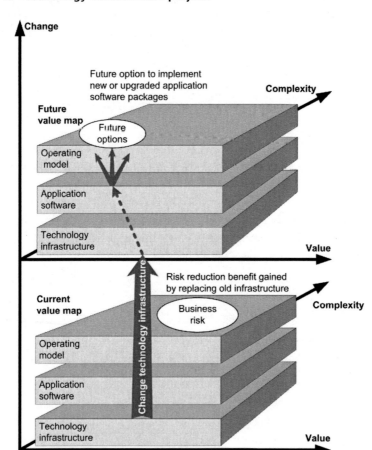

Building the technology infrastructure is a complex and multi-layered task. There is no direct connection to business needs or to the operating model: rather, the IT team must try and predict what applications will be needed in the future and design the infrastructure accordingly. The key consideration is adherence to standards – when a lower layer is built in accordance with a well-known standard, it will generally allow more options for the layer above. For example, many Windows applications are capable of running on several versions of the Windows operating system, which is very convenient for the technology team – they can simply install the latest version of Windows and be reasonably confident that application software that is designed for an earlier version will still run. This is called 'backwards compatibility' and is a very important weapon in the technology team's armoury.

As another example, when IT people build a network of computers, they will generally use a communications protocol called TCP/IP, which is the standard on which the internet is built. Below the TCP/IP layer, there might be a variety of supporting technologies. One business might have TCP/IP operating over a wireless network, while another uses physical cables. The TCP/IP layer needs to be capable of working with both these technologies, the wireless network and the cables. But the layers *above* TCP/IP do not need to know how to work with either wireless or cable networks, because these are not directly below them. For example, one of the layers above TCP/IP might be an internet browser, such as Microsoft's Internet Explorer. A browser has no need to know what is underneath the TCP/IP network on which it is running. Similarly, the TCP/IP layer has no 'understanding' of the browser that runs on it.

These rules help make the lives of IT people somewhat easier, because they can swap some of the elements of a technology stack without disturbing the layers above. In the example above, an IT engineer can change from cables to a wireless network without changing any of the software that runs on the network. In a sense, the TCP/IP layer provides a level of 'insulation' between the upper layers and the lower network.

Scale, meanwhile, has the effect of setting standards. Perhaps the best example is, again, the Microsoft Windows operating system; whatever the technical merits of Windows, it has the undeniable advantage of supporting many more applications than any other operating system. Thus a business that selects Windows as its operating system will give itself many more future options at the applications-software level.

So it is not just a question of reducing risk. Technology-infrastructure projects also generate future options for software applications, which may lead to future options at the business level. And many of these future options will be unknown when the project is implemented, because of the potential to use application software that has yet to be written.

Cost-reduction Programmes

Notwithstanding all the above, a large number of technology-infrastructure projects have only one aim: to cut direct IT costs. It is not suggested here that this ambition is inappropriate, or that it can be *replaced* by the focus on value and options set out above. Instead, the objective of cost-cutting should be seen in parallel with this broader perspective.

For example, Cisco Systems made a significant cut in the cost of its internal IT storage by moving away from traditional disc drives towards networked disk arrays and central storage devices. In doing so, it cut the cost per megabyte of storage by 75 per cent. This move would not have helped Cisco gain a competitive advantage, but it will have contributed to the broad objective of cost control.[102]

Figure 41 shows that the benefits of a cost-reduction programme are very limited. The business operating model and software applications remain much as they were.

Figure 41 – IT cost-cutting projects

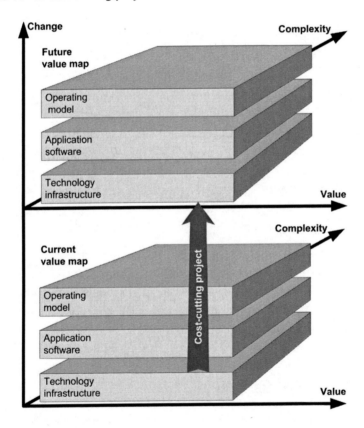

In cost-reduction programmes, the cash saving *is* the total benefit. Such a project generates no future options at all; the business does not move forward by so much as an inch; and it appears at the end of our list of project types because it is the only one where this holds true – all the others have a mixture of benefits. This means we have finally arrived at a type of IT project where the ROI approach can truly be said to give the total benefit. All that the project aims to do is cut the direct cost of the IT platform, which is something that can be estimated and fed into an ROI calculation. The costs can also be measured and recorded in the future, in order to confirm the validity of the ROI calculation.

Note that we are talking here about cutting *direct* IT costs. Many IT projects aim to reduce broader operational overheads (sales, general and administrative expense, or SG&A). For example, when a company sets up a shared service centre and centralises its teams for accounts receivable and payable, it is reducing its broader overheads. But that is not the same as cutting the direct cost of IT; indeed, in the short term, the direct IT costs might increase, for example as the result of setting up a shared service centre. From the perspective of our framework, this is actually an improvement project because the business operating model is being made more efficient. The investment in IT leads to a reduction in broader SG&A spending.

Similar comments might be made about outsourcing. It is fine for the IT infrastructure, but does not really work for the application layer, and particularly for those applications that sharpen a firm's competitive edge. Here, a business should keep control of the applications. But if the outsourcing project is carried out purely to cut direct IT costs, managers should be wary – such a project might be easily justified, but it will not help the business compete over the long term. A discussion on IT value might well start with cost-cutting, but it should not end there.

Prioritising Business as Usual and Building a Decision Table

We have talked a great deal about investing in IT projects, but have yet to discuss the biggest 'investment' of all – that of maintaining business-as-usual. For most firms, the IT operating expense ('opex') budget dwarfs the capital expense ('capex') budget. And the opex spend is almost all concerned with maintaining business-as-usual – with simply keeping things running.

Much of this opex spend is labour, and arguably much of it is misallocated. Experience tends to show that IT people actually spend most of their time dealing with new problems rather than maintaining old systems, and this is because software, as we have seen, does not really break. Maintaining a software system is fundamentally different to maintaining a car, because cars break down and software does not. Much of the licensing cost for maintaining an ERP system is actually about improvements and incremental upgrades, which should really be treated as capex.

How does the business-as-usual spend look in the framework? The answer is immediately apparent: the opex spending on business-as-usual is aimed at mitigating risk and maintaining a capability to take up future options. It is conceptually similar to technology upgrades, in that it generates no *immediate* business benefits – otherwise it would presumably be treated as programme cost; rather, it is aimed at keeping the current systems running. This is shown in Figure 42. Firms spend money in their IT opex budgets in order to manage risk and keep their options open. Again, the ROI toolkit breaks down completely for business-as-usual: there is no additional return on investment by simply keeping the business running.

Figure 42 – Business-as-usual (BAU) spending

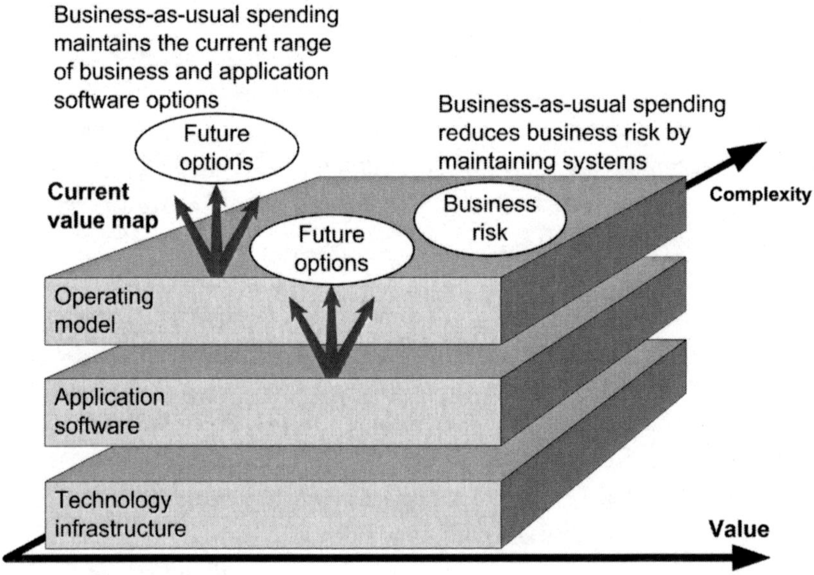

Finally, we can build a table to map the archetypal projects and IT opex spend against the benefits we have identified along the way. This is shown in Figure 43. This table is our second key deliverable, the first being the framework itself, and is a practical tool that can help managers make better decisions on IT projects.

Figure 43 – Mapping the projects onto an IT investment decision-table

Project type	Improved automation & information	Scaling the operating model	Reducing risk	Future options
Bespoke software	✔	✗	✗	✔
Packaged software	✔	✗	✗	✔
Technology-driven initiative	✔	✗	✗	✔
Technical upgrade	✗	✗	✔	✔
IT separation	✗	✔	✗	✗
Organic expansion project	✗	✔	✗	✗
Basic integration	✗	✔	✗	✗
Deep integration	✔	✔	✗	✔
ERP programmes	✔	✔	✗	✔
Infrastructure upgrades	✗	✗	✔	✔
IT cost-reduction projects	✗	✗	✗	✗
Business-as-usual opex spend	✗	✗	✔	✔

Part IV
Making Better IT Investment Decisions

In Part III we used the new framework to analyse a range of typical information-system projects. We found that many of these are actually business projects with a technology component. Others are about reducing risk, but also about creating options for the future, and many of the most complex projects involve changes to people and processes. The framework shows each type of project in a different way – there is no one single yardstick that can be used to assess all IT projects. Each has its own unique 'fingerprint', in the sense of the benefits that it is likely to deliver.

We finished our review of IT projects where many companies begin, with simple cost reduction. If nothing else, Part III showed that cost reduction is *not* the only benefit of IT projects. It was also demonstrated how business-as-usual spending can be shown in the framework, in a similar manner as for projects; this might allow for such spending to be prioritised against project spend.

The table (Figure 43) at the end of Part III is the second key deliverable of this book. It provides a simple tool for analysing a portfolio of projects, and debating their value. This basic decision-making table may prove sufficient for many managers' purposes. After all, one of the objectives of this book is to try and improve the debate between business and IT people, and the decision table should certainly help with this.

But some might want a more numerical approach. In this final part of the book, we will start to move towards just such a numerical model. We will also draw together some of the ideas we have discussed into recommendations for making better decisions about IT.

Chapter 11
A Pragmatic Alternative to ROI

A better title for this chapter might be 'A rough-and-ready alternative to ROI'. The framework offers no clear answers in the way that ROI does (or claims to). Instead, it offers a practical alternative to the assessment of IT projects. But surely a rough-and-ready tool that actually works is better than a supposedly precise measurement that fails to work?

First, a word of caution. Although it is possible to assign a value to the benefits and options created by an IT project, some fairly heroic assumptions about future value have to be made. And the future options have to be identified before they can start to be valued. It might be preferable to assess value and future options using the decision table presented in Part III, and leave it at that. But some managers may want more rigour behind their investment decisions, or at least a numerical rating of value. This final section sets out such an approach. I hesitated before writing this chapter, because my belief is that decisions on IT investments always, in the end, come down to the experience of the managers making them. But I decided to include it anyway as a means of closing the loop and re-connecting with ROI.

So how can managers decide between competing priorities for IT spending? Put another way, how can they make best use of a limited IT budget when they are comparing different types of project? Each type has a different goal, and each looks different in the framework, as we saw in Part III. Some projects have the goal of improving automation in a particular process; others are designed to improve information across the entire organisation; some infrastructure projects have no immediate benefits, but rather will lay the ground for future options. This problem is faced by every business that invests in IT systems.

Business Scope

Here is the first part of a pragmatic approach to prioritising IT projects. Let's start with application software, and make use of the new framework as a visual model. By doing this, we can visualise the impact of an application-software project by examining which parts of the operating model will be affected by it.

As an example, consider a new training portal which allows people to select the training courses that they would like to attend. This will potentially benefit everyone in the organisation. But now consider a new automation system in one manufacturing plant – the benefits will be felt in that plant alone. The 'business scope' of each project is completely different. Note that when talking about business scope we are *not* talking about scope in the normal sense of the word, as used in IT project management, for example. In that sense, the scope of the project is its boundaries. It forms part of the project definition, and indeed one of the key dangers to avoid in any project is 'scope creep'. But here we are using business scope in a very different way – we are talking about the scope of those parts of the business that will be *affected* by the applications-software project.

We can use the concept of business scope to help decide whether to approve an IT project, by treating its business scope as a key component of its value. The framework provides a convenient way of seeing the business scope visually. First, we look at a project with a narrow scope, such as the new automation system in a single plant, as shown in Figure 44.

Figure 44 – Application with a narrow business scope

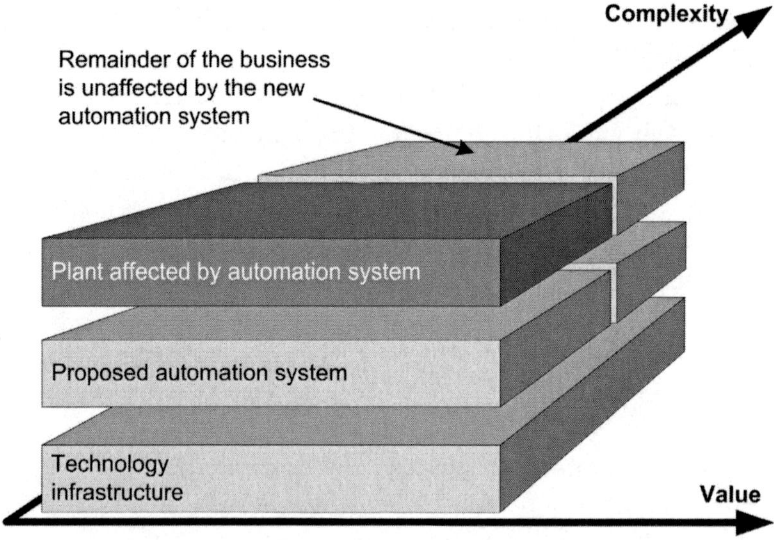

This visual approach provides a clear upper limit to the effects of each project. The automation system in one plant will only realistically affect the costs and revenues of that plant; but a new training portal can potentially affect people's productivity across the whole group, as shown in Figure 45.

Figure 45 – Application with a broad business scope

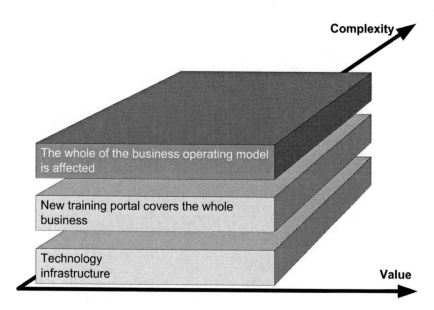

The whole of the business operating model is affected

New training portal covers the whole business

Technology infrastructure

Complexity

Value

If we want to compare projects, we need to express the business scope in financial terms, and we can do this by examining the revenues or costs that are associated with it. In the examples above, the automation system in one plant has a business scope that equates to the running costs of that plant. In the case of the training portal, the business scope encompasses the costs of running the whole group. Again, note that these are *not* the costs of implementing the IT project, but the business costs for that part of the operating model that will be affected by the project.

Whether business scope is measured in terms of costs or in terms of revenues is a matter of choice. Within a single business or business unit it is better to compare costs, as described in the example above, since the entities in the business operating model will be cost centres. But within a larger group of businesses it might be better to compare revenues, since that is the measure of each. Of course, both revenues and costs are denominated in the same unit of measurement (cash). In terms of business impact, reducing costs by one million is the same as increasing revenues by one million (ignoring any more complex effects, such as tax). So, with care, costs can be compared with revenues, since changes in each are measured in the same unit.

We look now at technology-infrastructure projects. In order to see the business scope of such a project, we can use the same basic idea. All we need to do is

ignore the applications-software level for a moment. First, let's look at a
network upgrade that affects only part of the group, as shown in Figure 46.

Figure 46 – Technology infrastructure project with a narrow scope

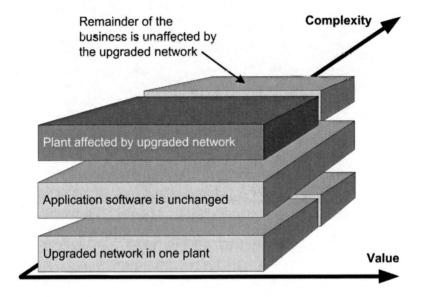

Next, let's look at an infrastructure project with a business scope that
encompasses the whole group; this is shown in Figure 47. Again, the point is
that the business scope of an infrastructure project can be assessed by ignoring
the application layer. (As an aside, it is interesting that many infrastructure
projects, such as an upgrade to the latest version of Windows, have group-
wide business scopes.)

Figure 47 – Technology-infrastructure project with a broad scope

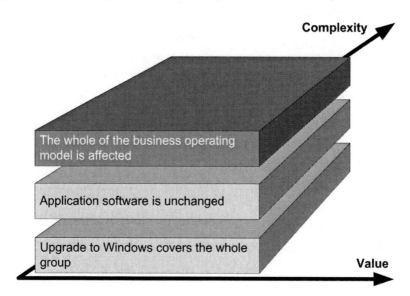

We have started to look at the value of an IT project in terms of business scope. But we clearly need to go further than this, because the effect of a project does not depend solely on the business scope.

Potential Impact

Even though a new automation system for a manufacturing plant has a limited business scope, it could still have a significant impact. For example, it might lead to an increase in throughput or a reduction in labour costs. On the other hand, a new training portal is likely only to make the company's existing training programmes more visible and accessible to staff – it will have little discernable effect on the bottom line. Even though the scope of the training portal is broader, the potential business impact is less. This concept of potential impact is part of a pragmatic alternative to ROI.

In order to arrive at an estimate of value, we need to convert the potential impact into financial terms. In this way we can make a *financial* distinction between a project with a broad business scope but light impact, such as an intranet, and one with a narrow business scope but a high impact, such as a plant-automation project. This top-down approach allows for different types of projects to be compared.

Let's start with the two sample projects we have already looked at. It might be estimated that a new automation system will reduce costs in a plant by 5 per cent. Conversely, a new training portal will have an effect on costs across the whole plant of (say) 0.1 per cent, by leading to better-trained staff and therefore improved efficiency. Assigning percentages to the effects has the advantage that we now have a number. The automation system will reduce plant costs by 5 per cent, which gives an absolute number, and the intranet will reduce group-wide costs by 0.1 per cent, which also gives an absolute number. The two numbers can now be compared to see which project makes more financial sense.

We can call this absolute number the 'notional value', notional because it will usually be unrelated to real cash flows. (This point is critically important, for reasons which will become clear.) In simple terms, we are defining notional value as:

Notional value = business scope x potential impact.

This might sound rather obvious, as indeed it is: in fact, many businesses effectively use this approach when they say that an application is 'strategic'. They are really saying that the business scope is broad and the potential impact high. Again, the concept of notional value is intended to help in making better decisions, not to produce numbers which can match an accountant's requirement for rigour. And at least we are starting from the right perspective, by working top-down from the business level. The framework suggests that every IT project should be evaluated in terms of its effect or potential effect on the business operating model.

But the obvious question is, how is the potential impact to be calculated? This is the difficult part, for which there is no magic formula. We have come full circle in our search for the true value of information technology, because we are back to calculating the value of the associated business improvements. The big difference is that we are starting from a top-down perspective, based on business scope and potential impact on the operating model.

It helps that we are trying to estimate value for a particular purpose: to help IT managers make better decisions about what projects to prioritise. It is vital that this question is assessed separately from any considerations about real cash flows. In this way the estimates do not need to be of the rigorous standard required for external scrutiny – they simply need to be good enough to make the right investment decisions.

But we can carry on a bit further and see if we can build at least *some* estimate of the notional value. Any estimate, after all, is better than none. The rationale for this approach comes partly from a book by Douglas Hubbard called *How to Measure Anything*.[103] He defines a measurement as anything that reduces uncertainty, and quotes a famous example from the physicist Enrico Fermi, who asked his students to estimate the number of piano tuners in Chicago:

His students – science and engineering majors – would begin by saying that they could not possibly know anything about such a quantity.[104]

But Fermi would press them to break the problem down and start to come up with estimates that, at least, reduced the uncertainty.

He would start by asking them to estimate other things about pianos and piano tuners that, while still uncertain, might seem easier to estimate. These included the current population of Chicago (a little over 3 million in the 1930s to 1950s), the average number of people per household (2 or 3), the share of households with regularly tuned pianos (not more than 1 in 10 but not less than 1 in 30), the required frequency of tuning (perhaps 1 a year, on average), how many pianos a tuner could tune in a day (4 or 5, including travel time), and how many days a year the tuner works (say, 250 or so). The results would be computed:

Tuners in Chicago = population / people per household

   x percentage of households with tuned pianos

   x tunings per year / (tunings per tuner per day x workdays per year)

Depending on the specific values you chose, you would probably get answers in the range of 20 to 200, with something around 50 being fairly common. When this number was compared to the actual number (which Fermi might get from the phone directory or a guild list), it was always closer to the true value than the students would have guessed.[105]

The statement that it is hard to measure the value of an IT system is true, but not particularly helpful. It is better to work out some upper and lower bounds on the value and then try to improve these. Just by reducing the uncertainty in the estimate of IT value, it is possible to make a better decision on whether to buy it. (Hubbard gives some compelling examples of how to reduce the uncertainty in a wide range of problems, not just in IT, and his book is strongly recommended.)

Although there is no magic formula to measure potential impact, one obvious conclusion comes straight out of the new framework. This is that the potential impact must be calculated in different ways for different types of benefit. In Part III, we saw that some IT projects increase revenues, others reduce costs, others give options, and some protect revenues that are already there. In the following four sections, I will attempt to sketch out ways that the notional value of each benefit type might be estimated.

Calculating Value From Improvement Projects

We start with the benefits that come from improving the business, and particularly the benefits that derive from automation. Here we can relate the potential impact of a project to underlying operational and financial levers, including the key operational objectives of speed, quality, flexibility, cost and dependability. Ultimately, automation projects should improve at least one of these: orders should be processed more quickly; quality should be improved through better feedback; the system should allow for greater customisation; it should be cheaper; or it should allow for higher dependability.

Businesses already know how to calculate these benefits. For simple automation projects, the approach set out here is not in fact markedly different from that of the standard ROI approach, since the benefits can be identified and calculated. This is the one area where the standard bottom-up ROI approach works pretty well, and we can use standard tools.

The UK water company Severn Trent implemented an interactive voice messaging (IVM) system in the call centre that handled its domestic-customer debt collections. It gained immediate efficiency benefits because the system filtered out wrong numbers before call-centre agents tried to call. Interviewed by *Computer Weekly*, Russell Mackuin, debt strategy and controls manager at Severn Trent, said:

Prior to using [IVM], up to 50–60% of the time a call centre agent would get through to someone other than the person responsible for payment, using manual dial out processes or predictive diallers, which can be incredibly hit or miss. Now, through IVM, Severn Trent can filter out wrong party numbers before the agent speaks to the party, making the agent more efficient.[106]

That type of benefit is easy enough to quantify, by measuring the time taken per call. Similar calculations can be made for productivity applications such as CAD systems. For example, the Australian airline Qantas found that design time was reduced by 40 per cent by implementing CAD and CAM (computer-aided manufacturing) systems.[107]

But sometimes IT spending is defensive – its aim is to protect existing revenues. In his book *Information Technology and the Productivity Paradox*, Henry Lucas set out to examine the real sources of value for IT systems. He created an 'IT Investment Opportunities Matrix' to show how different types of IT project add value. One of these types is what he calls 'no other choice' investments.[108] These are investments which a firm has to make just to stay in business. For example, a retail bank nowadays has to have a website that allows people to check balances and make simple transactions. Years ago, there might have been a business advantage for the first bank to do this; but that is long gone. Such online facilities are now expected by customers, as 'hygiene factors', so there is no increase in revenues associated with the investment, and no plausible ROI case. Nonetheless, we can still calculate the cost of *not* making the investment, on a revenue-protection basis. That is, how much market share would a retail bank lose by *not* having a transactional website? We might expect this number to be quite large in the first year, up to say 5 per cent, and for it to increase fairly rapidly in the future. On that basis, some useful estimates of percentage business-impact can be made. A similar point was made by Robert Kaplan when describing investments in computer-integrated manufacturing (CIM):

In fact, the correct alternative to new CIM investment should assume a situation of declining cash flows, market share, and profit margins. Once a valuable new process technology becomes available, even if one company decides not to invest in it, the likelihood is that some of its competitors will.[109]

Kaplan went on to quote Henry Ford: "If you need a new machine and don't buy it, you pay for it without getting it."[110] So we can often calculate the benefits of an automation project by estimating the negative effects of *not* doing the project. Note that this is hard to estimate using an ROI approach, since there are no additional returns to the company, but instead a reduction in potential losses.

Conversely, some projects give benefits in increased revenues, as opposed to reduced costs. We talked about Cathay Pacific earlier, with its scheme to allow people to check in to their flights using a mobile phone.[111] Is it too much to suggest that the airline will win new customers from this innovation? That some people, given a choice of airlines, will choose Cathay Pacific simply because it is easier to check in? If so, the benefit will be straightforward to estimate, in terms of new customers and their likely spend on flights. The difficulty here is not in fact in the calculation of value, but in finding such revenue-enhancing opportunities in the first place.

What about improving the business through better information? This will allow a firm to communicate better, to provide better management information or simply to act as a more cohesive holistic entity (information being the glue that binds a firm together). It is impossible to work out an exact number for the value of information, but for decision-making purposes there is no need to. As Hubbard said: "... a measurement doesn't have to eliminate uncertainty after all. A mere *reduction* in uncertainty counts as a measurement and possibly can be worth much more than the cost of the measurement."[112] Following Hubbard, we can ask: what would happen if you did not have this information? What decisions would you make differently, and what might that cost? He describes several ways in which information has value, but for business purposes the one that matters is: "Information reduces uncertainty about decisions that have economic consequences."[113]

Again, recall that we are trying to produce estimates for notional value that will allow an IT manager to prioritise the projects in his or her portfolio. An example of one such project in a retail business might be an IT reporting system that provides a summary of the value of stock across all stores and warehouses. What value would the management team place on having accurate stock figures across all their business? The answer is surely that it would help them to make better decisions on the quantities of stock to buy or produce. Here, the value of the information system is in preventing 'stock-outs', where sales are lost because the retail outlet is out of stock, and in reducing excessive inventory, which ties up cash. Since the probable costs of both can be calculated relatively easily, we have arrived at an estimate for the value of the IT reporting system: it is the likely reduction in these business costs by virtue of providing better information.

In summary, the benefits of automation projects can usually be worked out directly, using current methods. For revenue-protection projects, a 'what if' approach can be used to estimate value. And the benefits of information improvements can be derived by thinking about the true value of making better decisions.

Calculating the Benefits of Scaling the Business

Take the case of an IT manager whose company is selling one of its divisions: the manager has the task of separating off the division's IT system from the rest of the company's. This is a big IT project, and the books tell us that we need to build an ROI case for it, just as we must for any other project. But as we have already seen, this is simply not sensible – the ROI case for such an IT project cannot be made in isolation. The business is selling a division, and the value will be gained at the business level. The job of our hypothetical IT manager is to make the business project happen.

The notional value of the IT project is effectively the same as the value of the business project, which is effectively the value of the divested division. The project-related costs and revenues are those associated with the entity which is to be sold, and the potential impact of the separation project is 100 per cent. After all, if the separation programme is unsuccessful, the divested division cannot operate independently of the group that has sold it.

The same logic applies when a retail business opens a new store; for practical purposes, it might as well be assumed that the potential impact is 100 per cent, and that the notional value of the IT project is equal to the revenue projected for the new store. Of course, an attempt *could* be made to open a new store without IT systems; it might be argued that it could operate anyway, but that it would be rather hobbled in its daily operations – there would be no point-of-sale tills, no automatic inventory, no daily sales figures, and so on. Nonetheless, the store could still sell goods. So it *could* be said that the true value of the IT system is the difference between a store that operates normally and one that is severely constrained, and a figure of 50 per cent (or less) of forecast revenue might be assumed.

However, this will still produce a big number. For example, in Part III we looked at the demerger of the DaimlerChrysler group. The separation project for the financial services business had a cost of less than $50 million, but it allowed a business with a portfolio now worth $32 billion to be created.[114] In terms of setting priorities, the separation project was clearly worth a substantial fraction of that $32 billion.

It might be objected that these numbers are absurdly large. If the true value of DaimlerChrysler's separation project were really as large as $32 billion, it would justify *any* amount of IT project spending. But the whole point of estimating notional value is to help a CIO to prioritise the IT budget. The fact that this sort of 'business scale' project has a very high notional value reflects the reality that these projects are large-scale and high-priority (hence any CIO who fails to make a separation project happen is very likely to be fired). This does not justify allocating wildly profligate amounts of money to the project – the normal process of project specification and estimation will soon make clear what level of spend is actually required. But it does allow a CIO to put such a project at the top of his or her list of priorities.

Calculating the Benefits of Reducing Risk

When a company invests in an IT project to mitigate risk, it is really trying to protect its business and ultimately its revenues. This is not something that ROI can handle at all, because it is designed to deal with changes to existing cash flows, not with their protection. How then can we calculate the potential impact?

A different sort of calculation must be employed for risk mitigation. A good example is offered by a company that builds a back-up data centre. Here, the risk is of a major disaster affecting the business's information systems and data. The benefit of the new centre is obvious, and it can be calculated. All you need to do is work out the likelihood of a disaster striking, and the probable effect if it does.

Such calculations were put to the test in August 2005, when Hurricane Katrina struck New Orleans. Is it possible to calculate the likelihood of a hurricane striking the Gulf Coast, and the probable effects if one does? These calculations appear feasible, and would justify a significant spend on disaster-recovery provisions. Entergy, an energy company based in the city, was caught up in the hurricane, but had invested in a recovery plan to cope with such a situation, and during the hours and days that followed Katrina the company's team worked steadily to rebuild the IT systems, even when both the mains power and backup generator failed. The plan was successfully invoked, and at the same time Entergy was able to restore power quickly to its primary data centre.[115]

The probability of a disaster occurring can be calculated with a reasonable degree of accuracy, or at least plausibility, using insurance tables and the like. The potential impact can be estimated in terms of the time lost while the systems are unavailable, and the effect on market share of lost customers and damaged reputation. The beauty of notional value is that it can cope very well with such probabilistic thinking; it is effectively the expected value of a disaster occurring.

A similar calculation can arguably be applied to business-as-usual – the normal operating costs of running the information systems. Such spending is basically about keeping applications running, and minimising risk. Again, following Douglas Hubbard's prescription, we should estimate what would happen if we stopped spending money on supporting the information systems. This can quickly lead to a rough assessment of the potential impact in percentage terms, and quite frequently this will be large. If a business stops replacing its PCs, for example, they will eventually break down and people will be unable to do their jobs. A simple calculation of the probability that PCs will fail can be made, and a number arrived at for the effect on people's time.

Again, these calculations are imprecise, but they reflect the underlying reality of the decision to build a disaster-recovery centre, or to maintain spending on business-as-usual. In other words, the calculations might be complex as well as imprecise, but at least they are the right ones.

Calculating the Benefits of Options

To complete this numerical approach to prioritising projects, we need to work out the effects of future options. In order to value the options, we must first identify them. Then we need to derive a notional value, which is business scope multiplied by potential impact. Finally, we need to find a way to assess their potential *future* contribution to the *current* value of the project.

The first step is to identify the options, and this is no easy task. Installing a layer of software might give the business flexibility in the future, but in practice it is hard to work out what particular options will be enabled. But again, at least we are asking the right question. The real reason for installing the latest version of Windows across a business is to enable the option, in two years' time, of installing the latest applications software. At a minimum, it ought to be possible to pick out the key applications that might be required, even if these are simply upgrades to the current applications.

Assuming that we can identify at least some of the future options, we next need to value them, using some mixture of the three approaches we have already seen, depending on whether the options are for scale, future improvement or risk reduction. In other words, the future options will potentially enable future benefits. These benefits can be estimated in the same way as for planned benefits, as set out in the previous three sections.

Finally, we need to find a way to assess the current value of the future options. It might sound difficult to assign a value to an option which might or might not be taken up in the future (let alone to one which, as has been said above more than once, might not yet be discernable). For example, by installing the financial module of an ERP system, a business will generate a range of options for future modules, such as materials management and human resources. But how can we give these options a value when the business might not even exploit them? What is the value of a module that *might* be implemented?

Perhaps surprisingly, there is already a well-defined theory that can cope with these difficulties. The comparison has been widely made between flexibility in information-systems projects and options for financial stocks, using 'real options theory'. The comparison is shown in Figure 48, drawn from Henry Lucas's book *Information Technology and the Productivity Paradox*.

Figure 48 – Comparison of stock and IT project options[116]

	Call option on a stock	Real option on an IT project
Underlying asset	Common stock on which option is purchased	An IT system that will be developed in the future
Current value	Current price of stock	Expected present value of returns from the IT project
Volatility	Stock price fluctuations in the market	Uncertainty (variance) in expected cash flow from the IT project
Exercise price	Price in the option at which holder may buy shares when exercising the option	Investment required in IT project
Exercise date	Date on the option when it can be exercised	Date for developing the IT project

Using this comparison, we should be able to value the future options that are enabled by IT projects in the same way as stock options. A conventional method for evaluating the value of future options for stocks is the Black-Scholes model.[117] Can this model by applied to IT projects? A paper by Alfred Taudes, Markus Feurstein and Andreas Mild suggests that it can. The authors described an SAP upgrade at a central-European manufacturing company, estimating the value to the business of the project in terms of productivity improvements. So far, so good: this is the kind of straightforward automation benefits that we have already looked at. But the authors then also analysed the value of the *options* that were enabled by the SAP project. These were projects that simply could not have happened before SAP was upgraded. They included such possible projects as electronic data interchange (EDI), workflow and an e-commerce initiative. Using the Black-Scholes model, the authors were able to derive a value for the options. This value added to and exceeded the basic value of the SAP project. They had worked out an explicit value for the options.[118]

However, the analysis is complex, and this makes it hard to use the Black-Scholes method as a framework for debating IT value. There are also theoretical grounds for suggesting that the analogy between stock options and IT projects is not a good one, primarily because the latter are not tradeable, and thus have no resale value.[119]

As an alternative to the Black-Scholes calculation, the effects of future options on value can be estimated using a Monte Carlo simulation. To do this, we simply have to identify the future options and the 'probability' that they will happen. In the example above, the future options were EDI, workflow and an e-commerce initiative, and each had a notional value. The probability of each happening then needs to be estimated. (It is not in fact a question of probability but of management decisions – but this is the only way to make the approach work.) Finally, a computer is told to run a large number of simulations, with randomly selected options being exercised. Using this method, we arrive at a sort of 'expected value' for the future options.

This approach comes from a different theoretical direction to the real options theory. It does not explicitly allow for the value of being able to delay and change decisions, instead applying a simple probabilistic approach to the problem. In principle, this fails to reflect the underlying reality, because *probabilities* have to be assigned to future decisions, whereas in reality the latter will be under management's control. In the example above, management will *decide* whether to implement one or more of the options for EDI, workflow and an e-commerce initiative: there will be no simple roll of the dice. But the answers will probably not prove very different, and the Monte Carlo simulation approach has the advantage of being conceptually simple. It is also easier to calculate, and more importantly, easier to interpret. This makes it much more useful as a framework for debating value.[120]

But it has to be said that the valuation of options is complex, whether we use real options or the Monte Carlo approach. It is difficult to identify *exactly* what future options might be enabled by a particular IT investment, and then to work out their potential value. In many ways, it is easier to think about flexibility and leave it at that. Even if it is hard to measure the effect of options, we ought at least to be able to discuss them.

Building a Decision Table

We have finally arrived at a broad approach to prioritising IT projects. Our key objective was to establish a potential impact for each project or operating spend that can be used to establish its notional value. The sections above set out some ways of doing this for different benefits. (Recall that benefits are not the same as projects – in general a project will have multiple benefits.)

Now seems a good time to re-introduce the other side of the decision, the costs. Using this approach, we arrive at the modified table in Figure 49, where the business impact is estimated using the methods described above.

Figure 49 – Adding costs to the IT project decision-framework

Project	Notional Value / Potential Impact					Project costs
	Business scope	Improvement benefits	Scaling benefits	Risk-reduction benefits	Future options	
Overview of the project	Measured in either costs or revenues	Based on existing valuation tools and 'what if?' questions	Generally the total value of the associated business change	The expected value of the risks that are to be mitigated	Use either Black-Scholes or Monte Carlo	Assess using standard project-costing tools

Finally, we need to consider any direct IT savings. These are savings that will arise through a reduction in IT running costs. In Part I, it was said that it is a common mistake to focus too much on these potential direct IT savings, but this does not mean we should ignore them completely. Now that we have a table that shows the *real* value of IT projects, it is appropriate to consider such savings.

Rather than adding another column, it is easier to simply net off the IT savings against the project costs, and to call this the 'real cash effect'. We can call it 'real' because these costs will feed through into the company's accounts as cash – that is, we are assessing the net cash effect of the IT costs and savings to arrive at the total effect on the company's cash. In a sense, the suggestion is that we think of the direct IT savings as being 'negative costs'. This might sound like an accounting quibble, but the point is an important one – it is critical to separate the *benefits* of an IT project from the *real cash effects*, including any direct cash savings; this is the best way to avoid confusing benefits with cost reductions. This is not to say that cost control is not important; of course it is, but it cannot be the sole objective for the IT team. This modified decision table will look like Figure 50.

Figure 50 – Netting-off costs and direct IT savings

Project	Business scope	Notional Value / Potential Impact				Real cash effect
		Improvement benefits	Scaling benefits	Risk-reduction benefits	Future options	
Overview of the project	Measured in either costs or revenues	Based on existing valuation tools and 'what if?' questions	Generally the total value of the associated business change	The expected value of the risks that are to be mitigated	Use either Black-Scholes or Monte Carlo	Project costs netted off against direct IT savings

At this point, if we were to apply the appropriate discount factor to the real net cash effect, we would have arrived back at ROI, but based only on the direct IT costs. The big difference is that the ROI is now being evaluated in the context of the overall value of the project to the business. In other words, all the columns in Figure 50 are now being used together, instead of only the last one. From an accounting perspective, we would also need to consider the timescale over which we are assessing the costs and savings, whether over the first year, a total for the first five years, or the lifetime costs and savings. These are details for a management team to decide, but they should not obscure the main point – that we have finally arrived at a table which balances the *full value* of an IT project against the *full cash effect*.

Using the Decision Table

We have used the framework to identify the different ways in which IT can give benefits to a business, and have set out a decision table for businesses to use in assessing these benefits. Hopefully companies will build their own methodologies using some of these ideas. One pragmatic way of using the table might be to set a budget for IT as a whole, and then to manage the proposed IT projects within it. In other words, a business could ring-fence the IT budget and optimise spending within that, rather than trying to optimise its spending across all capital investments across the firm. There appears to be no clever way of replacing the (theoretical) capability of ROI to compare IT projects with other business initiatives; and, lacking that capability, it makes sense to optimise the IT budget in isolation. Luckily, this is a reasonably good fit with the way that many businesses manage their IT functions: they set a budget based on previous years' spending and industry benchmarks, and manage IT within that budget.

There might be a concern at this point about double-counting: it might appear that the table encourages managers to count value twice. For example, an applications project and an infrastructure project might appear independently, each with their own notional value, even though they are related and even though they support the same business scope. But there is in fact no double-counting, because these numbers are not real costs or revenues, but simply notional values. They are a convenient way to rank and assess business value, and they will not at any point be added together. The business depends on both projects, like links in a chain – if any one link is broken, the business suffers.

Incidentally, there is a hidden advantage to this new approach. By separating the real cash effect from the broader business benefits, a much freer assessment of these likely benefits can be made. The real cash effect can be rolled up into the company's financial projections, where it can be reviewed and audited, internally or externally. But the assessment of broader business benefits is strictly for internal management purposes – after all, no accountant will ever accept a value for future options. So management need not worry about justifying the numbers externally, but can simply use them for internal decision-making purposes.

Contrast this with the ROI approach, where managers have to try to identify benefits not in terms of notional value but as *real cash flows*. This is what the ROI approach really is: a full estimate of *all* future cash flows. These estimated future cash flows will inevitably get rolled up into the company's projections and internal budgets. And since the real cash flows are hard to identify, IT managers are understandably concerned that they will struggle to explain and defend their estimates. So, very frequently, they do not estimate them at all, but take an over-conservative perspective whereby the benefits are only alluded to and never calculated.

Partly, this conservatism is an inherent part of the budget process: it goes under the label 'sand-bagging', and allows managers room for manoeuvre. But what we are describing here is even worse than sand-bagging. Canny IT managers avoid getting drawn into a discussion of business benefits, because they know that any estimate might later be used against them. To avoid being held to a poor estimate of IT benefits, they would rather not assign them any value at all. This might be sensible from the perspective of conservative accounting, but it does nothing to help companies make good decisions on IT spending.

Assuming, however, that we have set a budget for IT, how do we next decide which projects to action within it? A simple way to prioritise the projects would be to list them in descending order of notional value. Interestingly, this approach will tend to militate against the sort of IT cost-cutting projects that are so prevalent. Assuming that managers start at the top and work downwards, they will presumably leave such projects to the end, since they have no notional value. And that would be a big step forward.

Chapter 12
Insights and Implications

In Part I of this book we discussed some of the difficulties and problems that are found with the ROI approach. This approach remains the default recommendation for capital investments, but is inappropriate for many, perhaps most, IT projects, because of the difficulty in calculating benefits. The framework presented in this book is a pragmatic alternative for making decisions about real IT projects.

Difficulties in calculating benefits are, at bottom, a result of the fundamental role of IT in supporting a business while not being part of the core value chain. It is simply not realistic to compare an investment in an IT system directly with a similar investment in a capital project that is part of that value chain. The IT system will give benefits over the long term and by generating future capabilities that might never be used. By contrast, an investment in a capital project that is part of the value chain may well generate clearly identifiable and measurable cash flows. The temptation is to invest in such projects at the expense of the IT systems, but that is not a route to long-term success.

One of the insights offered by the model is that business and IT strategies are not necessarily closely aligned. It will always be difficult to align them because so much of what IT offers is to be found in future capabilities and options. That might not sound particularly helpful to a management team needing to make decisions, but it is important to understand the limits of what can be planned. Once the business strategy has been operationalised, and those elements of it that change the operating model have been identified, then the IT *plans* (as opposed to strategy) can be designed. They can and should be detailed plans by that point, because the strategy is well defined.

Another insight is the importance of organisational complexity in making IT decisions. The reality of most businesses is that they struggle continually to build a single IT platform across the group. This is a never-ending battle. Businesses grow and shrink, acquire other businesses and then divest them, and the IT team must try to match the underlying systems in the context of this continual re-shaping of the corporate map. To ignore this factor, as most current approaches do, is to miss one of the key drivers behind IT projects.

One of the central messages that this book offers is that different IT projects offer very different benefits, and it is simply not sensible to try and assess them all using a simplistic ROI yardstick. An investment in a new version of Windows will offer almost no *immediate* benefits for most businesses. The real benefit is that the company will be able to use applications software that it *might* want in the future. By contrast, an investment in a bespoke software solution *will* offer immediate benefits, because it should support changes to a company's operating model.

To the question 'Is IT a strategic resource or simply a commodity?', the answer is that both labels are partly justifiable. The extreme view that IT is necessarily a strategic enabler appears to be wrong, but so is Nicholas Carr's view that it is a commodity, like electricity. A central theme is that IT strategy is all about applications, and an allied theme is that it is *not* about alignment with business plans. Much of the confusion lies in treating IT as a single homogeneous entity; the framework presented here offers a simple but powerful split between IT applications and IT infrastructure. Much of the latter can in fact be treated as a commodity, but this does not necessarily hold true at the application level, where a strategic advantage *can* be gained from close interaction between people, processes and technology.

The framework proposed here helps to resolve another paradox in the world of IT. How is it that spending on information technology is a relatively small part of most firms' budgets, but that IT plays such a large part in the real-world experience of these firms? A focus on costs gives a misleading impression of IT's importance, precisely because the costs are not all that high. The framework helps to indicate that, irrespective of costs, the IT platform has significant and ongoing effects on the business and on the future decisions it may take. Even a small IT investment can go a long way and last a long time, as the continued usage of mainframes shows. Such investments have long since been written down in accounting terms, yet their effects remain. The focus on costs is a fundamental part of our accounting framework, but it is of little help in making long-term IT decisions.

Plentiful examples have been given of where CIOs are doing the right thing according to this model, because real-world practices are clearly ahead of theory in this area. But what about the counter-examples? Are there areas where businesses are *not* making good decisions, according to the framework?

First, there may be a lack of focus on future capabilities. This book has developed the argument that IT investments should not just aim at helping the business now, but will also create capabilities that can be exploited in the future. This is especially important when it comes to the technology infrastructure, because so much of the investment that is made at this level is aimed at building future capabilities. The decision table set out in the previous chapter (Figure 50) may provide a means by which to emphasise the importance of future options in investment decisions; but it may be equally useful for managers simply to keep in mind the layered structure of IT, and to think about *flexibility* as a key objective in itself.

A second area of poor decision-making is an excessive focus on achieving a reduction in IT costs. We have touched on the question of IT costs throughout the book, starting with the suggestion that too great a focus on costs will lead managers to make bad decisions. The simple two-dimensional model illustrates this risk. It shows that while it is certainly *possible* to maintain the current systems, to demand that they be 'fit for purpose' and to pressurise the IT team to drive down their costs, such a cost-centric attitude will not help firms to realise the true potential of IT to improve their businesses. Those that see IT as a static support function will only ever have one lever to pull, and that is the cost-reduction lever. Managers need to build capabilities in their IT platform that will give them future options to take on the competition and win. Such capabilities cannot be built if the business is bent only on cutting costs.

Of course, when a business operates in a mature market, the strategy must often be for a continual reduction in costs. If a firm's markets are growing slowly or even shrinking, the only way to increase revenues is to take market share away from competitors, which is usually difficult. So the most practical way to improve profits is to cut costs through operational efficiency. The argument here is not against cost-cutting *per se*, but rather against a narrow focus on cutting the direct costs of the IT platform. At a minimum, managers should also think about using IT as a mechanism to cut broader operational costs, through better automation and information.

A third problem is that of seeing IT simply as a platform. With this mind-set, the objectives of the IT team are simply to manage a set of processes as efficiently as possible, according to a set of service level agreements (SLAs) that have been agreed with the business. Although firms that adopt this view generally understand that IT is important as an agent of change, this tends not to form part of their perspective when planning investments in technology. Instead, they are driven by short-term considerations: upgrading software because they have to, integrating systems in order to reduce complexity, and managing costs downwards.

But such businesses miss a critical element of IT's real role as a change agent, and the importance of using IT innovation to make continual improvements in the firm's operating model. A CIO needs to have two objectives: to manage the current IT systems efficiently, and to search for new opportunities for IT-driven business initiatives. Many IT departments are familiar with the first objective but need to understand and build their capabilities to achieve the second. To an extent, this is handled as a process of 'demand management', whereby the IT department will engage with the business, listen to its requests, and manage and prioritise them appropriately. However, this relegates the IT team to a relatively passive role, since they must rely on business people to spot new opportunities. More value can be added when IT people start to think themselves about how technology might shift the business forward in the future.

Another common tendency is to ignore or under-value the benefits of improved information (as opposed to automation). It is hard to put a financial number on the value of information. What, for instance, is the value of having one consistent picture across a business that lets the management team treat it as a single entity? There might not be an easy answer, but probably many businesses are simply ignoring the question. In the previous chapter, it was suggested that various thought experiments might help to clarify information benefits. But at a minimum companies need at least to acknowledge the existence of these benefits, and not simply ignore the value of information because it is hard to measure.

An allied problem was also described in the previous chapter. Often, managers do not even discuss the possible benefits of IT investments because they are concerned about being held to account for potential cash savings in the future, and about these potential cash savings being included within a company's forecasts and projections. There are probably many companies facing this hidden problem. The answer is surely to separate the estimation of value from the assessment of future cash projections. This is not a recipe for dishonesty – what is being suggested is simply that managers can acknowledge the existence of benefits from IT investments without necessarily seeking to assign those benefits a cash value, which then goes into the company's projections. It would be a better state of affairs if there were no conflict between estimating value and setting budgets, but there is, and this conflict is as much about the limitations of our current systems of accounting as it is about information technology. Managers need to distinguish between IT benefits and business benefits. Lower costs are not really business benefits; they are just lower costs. The decision table (Figure 50) set out in the previous chapter should be helpful in this respect, by explicitly separating the cash effect (including direct IT savings) from the broader business benefits.

Another problem in current practice is having the head of IT report to the CFO (Chief Financial Officer). This came about during a time when the accounting system was the most sophisticated software that a business might use. But it no longer makes sense: the framework clearly indicates the close linkages that exist between the business operating model and the IT systems. Given the omnipresence of IT within most companies' operations, it surely makes more sense for the head of IT to report to the COO (Chief Operations Officer), or for the two positions to be combined, as HSBC has done.

Companies struggle to see the value of investing in IT, and part of the reason can be found in the project-approval and gating process. An easy change would be very effective for many companies: simply to 'tick the box' when a business project can be actioned without any accompanying IT changes. That would help managers understand the layered and capability-building nature of IT investments. Every time a project has *no* need of any IT input, managers should acknowledge their own (or their predecessors') decisions to create the current IT platform, which has enabled the proposed business project.

And finally, the framework has implications for the influence of M&A activities on IT value. The framework shows that such value is built over time by creating capabilities and deploying them later. And the value of IT systems is closely bound up in their interactions with people and processes, and how these develop over time. When a company changes hands, however, much of this delicate balance is upset. Think about what happens when two businesses, each with its own long-term IT development programmes, decide to merge. Very often, the answer is that one of the programmes will be stopped in the interests of saving costs. The accumulated value and shared understanding that has gone into the programme is lost. A tentative conclusion may be drawn that M&A activities tend to destroy IT value, but a broader conclusion is that IT development programmes should be considered as part of any M&A deal.

Having looked at some of the fault lines in current practice, how do the framework and ideas presented here connect with the broader approaches and techniques of IT management? It was said at the start that this book is only about the value of IT systems, and that there has been no attempt to cover the whole range of issues that face an IT team. What comes out of the framework is, hopefully, a re-appraisal of the IT project portfolio, with a far greater emphasis on building future capabilities and a greatly reduced focus on cost-cutting. The projects that make up the portfolio need to be defined, analysed and costed; resources need to be assigned to them; they need to be managed to completion; and the day-to-day delivery of IT services needs to continue according to the service levels agreed with the business. This book does not cover these issues at all, but simply helps a company to define what systems it will build in the first place. That seemed to be a sufficiently complex problem to try and resolve.

Conclusions

This book was written to help resolve an old problem, the measurement of value in information systems. I hope it goes some way towards that objective.

Much has been written about whether information systems can confer a competitive advantage on a business, and even more about whether such an advantage can be sustained. The extreme perspective provided by Nicholas Carr says that no company can gain an advantage from such systems, because they have become a commodity which anyone can copy. In this view, any advantage that one business might gain from a new computer system can quickly be copied by its competitors, leaving them all back at square one. For example, the first bank that put in a network of automatic teller machines (ATMs) might have gained an edge over its competition, but nowadays these are a necessary part of doing business for any retail bank. They have changed from being a competitive advantage to being simply a cost.

However, this view probably overstates the case; and even if it were true, information systems would still be a necessary investment simply for businesses to survive. And the benefits to customers (or society in general, perhaps) would still be present. Although it might be true that no bank gains any benefit from its network of ATMs, it is certainly true that we all gain as customers by having more convenient access to our cash. The benefits might flow to customers instead of to businesses, but they are nonetheless there. In fact, Carr's points could be seen as a criticism of competition, or even of capitalism: arguably, *no* innovation can be protected forever, whether or not related to IT. The lengths to which pharmaceutical companies go to protect for as long as possible the innovations contained in their expensively-developed drugs are witness to this.

As we have noted, Carr's points about IT in general are really applicable to technology infrastructure only. It is probably fair to say that this infrastructure can be treated as a commodity, that it can be freely purchased, and that it confers no competitive advantage. Carr refers to another type of technology as being 'proprietary', in that it allows for a competitive advantage, but does not classify information technology as such.[121] But the applications-software layer interacts directly with a business's people and processes, which are constantly changing under competitive pressure. That gives an opportunity to a business to gain at least a temporary competitive advantage from the applications layer.

It might seem obvious, but an improvement in efficiency will not necessarily produce an increase in profit, because companies exist in a competitive environment within their marketplaces. Even when a company improves the *absolute* efficiency of its operations, it still needs to be significantly better *relative to its competitors* if it wants to convert that improved efficiency into higher profits. And there are many other factors which can affect a firm's profits, chief among them the intensity of competition in its sector. The net effect on profits can easily be zero if a company's competitors increase their level of efficiency at the same rate as the company does. In that case all the benefits of the improved efficiencies go to the firm's customers, in the form of reduced prices, higher quality, or more reliable deliveries.

Let's conclude with a question. Is it important to talk about investments in IT? I think it is, because information systems can enhance the quality of people's lives. This enhancement can be direct, for example in the recent innovations in banking by mobile telephone, which have made a difference to people's lives in Africa.[122] Or the enhancements can be indirect, for example in better productivity, which are really the improvements discussed in this book. When information technology systems are built right, they help businesses compete better through improved operational efficiency; and this means lower prices for consumers, or improved quality, or both.

It should also translate into improved productivity at the national level. This issue was reviewed by Erik Brynjolfsson and Adam Saunders;[123] they examined the evidence that IT investment by firms was related to increased productivity in the USA. From 1973 to 1995, US labour productivity increased by 1.4 per cent per year, but from 1996 to 2000 it increased by 2.6 per cent. Brynjolfsson and Saunders note that: "There is widespread agreement about the cause of this surge in productivity growth: information technology."[124] Productivity growth then increased further in 2001 to 2003, to 3.6 per cent per year. According to Brynjolfsson and Saunders, there is less consensus about the link between this further increase in productivity growth and investments in information technology. However, they went on to say: "Our belief is that the more recent surge is the result of IT, but in the form of a 'reap and harvest' story", when the US economy was benefiting from a surge in IT investment in the late 1990s.[125] They hypothesise that later trends in productivity growth can also be explained by variations in IT spending, but with a lag of approximately three to four years.[126] The framework set out in this book would be consistent with that theory, because of its emphasis on the mechanism by which IT spending creates future options that are only later converted into changes to firms' operating models.

Improved productivity at the national level is the key ingredient of improved living standards. There is a strong correlation between improvements in productivity and improvements in many other key indicators, including income, longevity and health. It could be argued that productivity will naturally increase as average wealth per capita increases. From a purely arithmetical perspective, this is no doubt true: a country's average wealth can be taken as gross domestic product (GDP) divided by the whole population, while productivity is roughly GDP divided by the working population. Assuming that the proportion of people in work does not vary significantly between countries, or over time in the same country, there will naturally be a close correlation between average wealth and productivity.

But there appears to be a stronger connection, and perhaps even a causal link. In his book *The Power of Productivity*, William Lewis sets out the results of decades of research, across many countries and industrial sectors, which aimed to understand how it is that some countries have been able to increase the living standards of their people, while others have failed to do so. His results suggest that improvements in productivity actually *drive* increases in average wealth and ultimately improvements in other life factors. He also found that productivity is best improved by encouraging competition between businesses.[127]

From an economic perspective, improvements in productivity are the best way to expand an economy. There are other ways of achieving economic growth, for example by managing a lower exchange rate or by increasing government spending. But they all have negative effects or risks, such as inflation, the crowding-out of private-sector investment, or the creation of a bloated state sector. By contrast, productivity gains have few, if any, long-term side effects; they reflect the gradual, incremental improvements that can be made on a daily basis.

The competitive pressure that all companies feel will never disappear: it is the most fundamental and tangible manifestation of a free market. Companies *must* innovate continuously in a competitive market, otherwise their competitors will do so and forge ahead. The continuous pressure for refinement in processes is a fundamental part of competition in a free market. This was expressed most clearly in 1776, the year that Adam Smith published his *Wealth of Nations*. He described the specialised jobs to be found in a pin factory of the day, and the improvement that specialisation offered over each person making the whole pin on his own.

One man draws out the wire; another straights it; a third cuts it; a fourth points it; a fifth grinds it at the top for receiving the head; to make the head requires two or three distinct operations; to put it on is a particular business; to whiten the pins is another; it is even a trade by itself to put them into a paper; and the important business of making a pin is, in this manner, divided into about eighteen distinct operations, which, in some manufactories, are all performed by distinct hands, though in others the same man will sometimes perform two or three of them.[128]

Smith went on to note the advantages in productivity that this specialisation gave: one man working alone could produce somewhere between one and 20 pins a day, but through specialisation each was able to produce an average of 4,800 pins a day. Already, more than 200 years ago, the process of continuous improvement, driven by competition, had begun.

Much later, in 1942, the economist Joseph Schumpeter described this continuous improvement as a process of 'creative destruction'. In his words: "Capitalism, then, is by nature a form or method of economic change and not only never is but never can be stationary." He went on:

The opening up of new markets, foreign or domestic, and the organizational development from the craft shop and factory to such concerns as U.S. Steel illustrate the same process of industrial mutation – if I may use the biological term – that incessantly revolutionizes the economic structure from within, incessantly destroying the old one, incessantly creating a new one. This process of Creative Destruction is the essential fact about capitalism. It is what capitalism consists in and what every capitalist concern has got to live in.[129]

He saw capitalism as the continual destruction of outdated ways of making products or delivering services, while replacing them with new methods. In the process, inefficient companies are replaced by strong and healthy new ones. Until the 20th century, these improvements were mainly limited to the physical production of goods. The technical innovations that created the automatic loom and the steam engine replaced the arduous manual labour of the farm and factory worker. But in modern economies, with a high proportion of service industries, one of the primary means by which creative destruction operates is through information technology.

We need to ensure as a society that people are sufficiently well educated and flexible to be capable of taking advantage of the new opportunities that are offered, at the same time as old jobs are destroyed. Technology can be a force for good, as long as we are flexible enough to adapt to it.

Notes

[1] For example, see Atrill, P. and McLaney, E. (2004) *Accounting and Finance for Non-Specialists*. 4th edn. Harlow: Pearson Education, pp. 282–317.

[2] Strassmann, P.A. (1997) *The Squandered Computer*. New Haven, CT: Information Economics Press, p. 5.

[3] Hamblen, M. (2006) 'Focus on ROI too limiting, Intel CIO says'. *Computerworld*, 29 May.

[4] e-Skills UK quarterly ICT enquiry 2006, quoted in McCue, A. (2006) 'Most companies "guess" tech ROI'. *BusinessWeek*, 24 May.

[5] Hoque, F. (2002) *The Alignment Effect: How to Get Real Business Value out of Technology*. Upper Saddle River, NJ: Financial Times/Prentice Hall, p. 50.

[6] Strassmann, *The Squandered Computer*, p. 173.

[7] Ibid, p. 172.

[8] PRINCE2 is a registered trademark of the Office of Government Commerce (OGC) in the UK. Information obtained from the OGC website at *www.ogc.gov.uk/methods_prince_2.asp*, accessed 30 December 2009.

[9] Porter, M.E. (1985) *Competitive Advantage*. New York: Free Press.

[10] Adapted from ibid.

[11] Carr, N.G. (2003) 'IT Doesn't Matter'. *Harvard Business Review*, Vol. 81, No. 5, pp. 41–50.

[12] A related point is made in Mata, F.J., Fuerst, W.L. and Barney, J.B. (1995) 'Information Technology and Sustained Competitive Advantage: A Resource-Based Analysis'. *MIS Quarterly*, Vol. 19, No. 4, pp. 487–505. The authors identify the ability of managers to build applications that support other business functions as being the only sustainable source of competitive advantage that can be gained from IT.

[13] Data from *Financial Times Global 500*, 2008.

[14] Wailgum, T. (2009) 'GE CIO gets his head in the cloud for new SaaS supply chain app'. *CIO*, 22 January.

[15] Slywotzky, A. and Morrison, D.J. (2000) 'Concrete solution – company operations'. *Industry Standard*, 28 August.

[16] Hadfield, W. (2005) 'Whitbread integrates hotel systems for £2.5m'. *ComputerWeekly*, 15 November.

[17] Weier, M.H. (2007) 'SAP rollout doesn't come easy for Kimberly-Clark'. *InformationWeek*, 21 June.

[18] Information on the Windows API obtained from Microsoft Corporation's website, at *msdn.microsoft.com/en-us/library/aa383750(VS.85).aspx*, accessed 30 December 2009.

[19] General Electric annual report, 2008.

[20] De Beers's operating and financial review, 2008, p. 1.

[21] Adapted from ibid.

[22] I have found many references to the phrase 'value map', but none that claim a copyright of this phrase. If anyone holds such a copyright, I will be happy to acknowledge it.

[23] Hammer, M. (1996) *Beyond Reengineering*. New York: HarperCollins, p. 13.

[24] Harris Corporation annual report, 2009.

[25] Travis, P. (2005) 'Supply-Chain Engineering: Harris is taking a novel approach to its supply chain by upgrading IT at the start of the production process'. *InformationWeek*, 30 May.

[26] Yetton, P.W., Johnston, K.D. and Craig, J.F. (1994) 'Computer-Aided Architects: A Case Study of IT and Strategic Change'. *MIT Sloan Management Review*, Vol. 35, No. 4, (Summer), pp. 57–67.

[27] Grant, I. (2008) 'Swisscom efficiency peaks with Motorola PDAs'. *ComputerWeekly*, 14 November.

[28] Drucker, P.F. (1979) *Management*. London: Pan Books, p. 57.

[29] Lafley, A.G. and Charan, R. (2008) *The Game-Changer: How Every Leader Can Drive Everyday Innovation*. London: Profile Books, p. 9.

[30] Koch, C. (2007) 'IT's role in collaboration and innovation at Procter & Gamble'. *CIO*, 1 February.

[31] Information on Amazon's web services obtained from *aws.amazon.com*, accessed on 30 December 2009.

[32] Gordon, D.T. (2001) 'John Hill aims to excite Praxair IT staff'. *CIO*, 15 May.

[33] *The Economist*, 10–16 October 2009, pp. 73–4.

[34] Dignan, L. (2003) 'Clorox brightens its IT with SAP'. *Baseline Magazine*, 8 August.

[35] Manenti, P., Parker, B. and Veronesi, L. (2009) 'SaaS ERP system with Oracle at Farwest Steel Corporation: a case study'. *Manufacturing Insights*, January.

[36] Cowley, S. (2005) 'JetBlue picks SAP for major ERP rollout'. *Infoworld*, 11 May.

[37] O'Donnell, A. (2009) 'Assurant's VerifyIns site drives online sales success'. *Insurance and Technology*, 21 January.

[38] A related point is made in Soh, C. and Markus, M.L. (1995) 'How IT creates value: A process theory synthesis'. *Proceedings of the 16th Annual Conference on Information Systems*, Amsterdam: The Netherlands, pp. 29–41. The authors set out a three-stage model for converting IT into business value. Under this model, IT spending must first create IT assets, which must then be used to create IT impacts. Only then can IT impacts be converted into organisational performance.

[39] UPS website, at *www.ups.com/content/gb/en/about/facts/index.html*, accessed on 26 August 2009.

[40] Ross, J.W. (2001) *United Parcel Services: Delivering Packages and E-Commerce Solutions*. MIT Sloan School of Management Working Paper No. 4356-01.

[41] Keynote address, *Redefining IT's Value to the Enterprise*, given by David Barnes, CIO at UPS, to the Forrester IT Forum in Las Vegas, 19 May 2009. UPS website, at *www.pressroom.ups.com/About+UPS/UPS+Leadership/Speeches/David+Barnes/Redefining+IT percent27s+Value+to+the+Enterprise*, accessed on 26 August 2009.

[42] *The Laureate, Journal of the Computerworld Information Technology Awards Foundation*, June 2009, pp 81–8.

[43] Quoted in Lucas, H.C. (1999) *Information Technology and the Productivity Paradox: Assessing the Value of Investing in IT*. New York: Oxford University Press, p. 50.

[44] McDougall, P. (2006) 'Lufthansa IT chief looks to cut costs with SOA software'. *InformationWeek*, 21 November.

[45] Brynjolfsson, E. and Hitt, L. (2003) 'Computing productivity: firm-level evidence'. *Review of Economics and Statistics*, Vol. 85, No. 4, pp. 793–808.

[46] Brynjolfsson, E. and Saunders, A. (2010) *Wired for Innovation: How Information Technology Is Reshaping the Economy*. Cambridge, MA: MIT Press, pp. 70–1.

[47] Mintzberg, H. (1996) 'Reply to Michael Goold'. *California Management Review*, Vol. 38, No. 4, pp. 96–9.

[48] Mark, R. (2009) 'American expands domestic wi-fi service'. *Eweek*, 31 March.

[49] SAP AG (2004) 'Consumer product companies run SAP'.

[50] Ho, V. (2008) 'Keeping up with fashion'. *ZDNet Asia*, 16 December.

[51] Strassmann, *The Squandered Computer*, p. 3.

[52] Society for Information Management (2008) 'SIM study: IT-business alignment continues to be top concern for IT executives'.

[53] McLaughlin, L. (2007) 'How IT shapes 'Store of the Future' for retailer family dollar'. *CIO*, 14 November.

[54] Bowie, M. (2003) 'Deutsche to add lanes to autobahn'. *RiskNews*, 31 March.

[55] *McKinsey Quarterly*, July 2009, at *www.mckinseyquarterly.com/What_health_systems_can_learn_from_Kaiser_Permanente_An_interview_with_Hal_Wolf_2397, accessed on 22 May 2010*.

[56] Datz, T. (2003) 'Merrill Lynch's billion dollar bet'. *CIO*, 15 September.

[57] Greengard, S. (2008) 'Chubb insures customer satisfaction with collaboration'. *Baseline Magazine*, 26 November.

[58] Daniel, D. (2007) 'Delivering customer happiness through operational business intelligence'. *CIO*, 6 December.

[59] 'John Deere starts up five-plant response-management deployment'. *Manufacturing Business Technology*, 6 November 2007.

[60] Southwest Airlines annual report 2008.

[61] Kontzer, T. (2005) 'Wings of change'. *InformationWeek*, 28 March.

[62] Steinert-Threlkeld, T. (2005) 'Southwest Airlines: flying low'. *Baseline Magazine*, 10 April.

[63] Babcock, C. (2005) 'How three midsize companies use business intelligence to their advantage'. *InformationWeek*, 11 May.

[64] Cisco website, at *www.cisco.com/web/about/ciscoitatwork/business_of_it/ERP_manufacturing_and_finance_web.html*, accessed on 11 April 2009.

[65] Knott, T. (2003) 'Computing colossus'. *Frontiers* (BP magazine), April.

[66] Patton, S. (2004) 'Management strategies: BP – in sync with his CEO'. *CIO*, 15 April.

[67] Ko, C. (2009) 'Check-in on Cathay Pacific with mobile phones'. *MIS-Asia*, 22 January.

[68] DaimlerChrysler AG annual report, 1998.

[69] Koch, C. (2003) 'CIO of DaimlerChrysler: Sue Unger's drive to diversity'. *CIO*, 15 June.

[70] Finkelstein, S. (2002) *The DaimlerChrysler Merger*. Case study no. 1-0071, Tuck School of Business, Dartmouth College.

[71] Daimler AG annual report, 2007.

[72] Weier, M.H. (2008) 'Daimler takes DIY to the extreme to extract financial systems from Chrysler'. *InformationWeek*, 18 October.

[73] Information from Specsavers's website, at *www.specsavers.co.uk/about/history*, accessed on 24 December 2009.

[74] Lewis, R. (2009) 'Specsavers goes global: CIO Michel Khan explains his vision'. *CIO*, 15 January.

[75] Duvall, M. (2003) 'Bank of America: three's a crowd?' *Baseline Magazine*, 5 December.

[76] Koch, 'CIO of DaimlerChrysler'.

[77] Worthen, B. (2007) 'How Coty tackled post-merger supply chain integration'. *CIO*, 15 January.

[78] Sullivan, L. (2005) 'Q&A with Disney's Patrizio: digital content is the future'. *InformationWeek*, 4 April.

[79] Schuman, E. (2005) 'Wawa CIO: upgrade fear dictated multimillion-dollar SAP purchase. *CIO Insight*, 20 May.

[80] Ross, J.W., Weill, P. and Robertson, D.C. (2006) *Enterprise Architecture as Strategy: Creating a Foundation for Business Execution*. Boston, MA: Harvard Business School Press, p. 29.

[81] Coase, R. (1937) 'The nature of the firm'. *Economica*, Vol. 4, No. 16, pp. 386–405.

[82] Parker, K. and Smith, F. (2007) 'To reach desired "IT end-state", Volvo CE focuses on multi-plant performance'. *Manufacturing Business Technology*, 1 October.

[83] Grant, I. (2008) 'Virtualisation is AB Foods' cuppa'. *ComputerWeekly*, 26 November.

[84] 2009 HSBC Holdings plc Interim Report: HSBC Annual Report 2008.

[85] HSBC investor presentation 'Take IT to the bank', given by Ken Harvey, CIO, 28 September 2006. Reviewed on HSBC's website at *www.hsbc.com/1/PA_1_1_S5/content/assets/investor_relations/060928_take _it_to_the_bank.pdf*, accessed on 26 August 2009.

[86] HSBC investor-relations event 'Becoming one HSBC', given by Ken Harvey, Group MD and Group Chief Technology and Services Officer, 28 October 2008. Webcast reviewed on HSBC's website at *www.hsbc.com/1/2/investor-relations//presentations-webcasts/investor-relations-event-becoming-one-hsb c#top*, accessed on 26 August 2009.

[87] Leimbach, T. (2008) 'The SAP story: evolution of SAP within the German software industry'. *IEEE Annals of the History of Computing*, IEEE Computer Society.

[88] Grant, I. (2008) 'Unilever credits SAP for boosting growth'. *ComputerWeekly*, 30 October.

[89] Sliwa, C. (2002) 'IT difficulties help take Kmart down'. *Computerworld*, 28 January.

[90] Koch, C. (2002) 'Supply chain: Hershey's bittersweet lesson'. *CIO*, 15 November.

[91] Carr, D.F. (2002) 'Hershey's sweet victory'. *Baseline Magazine*, 16 December.

[92] Bulkeley, W.M. (1996) 'When things go wrong: FoxMeyer Drug took a huge high-tech gamble; it didn't work'. *Wall Street Journal*, 18 November.

[93] Wailgum, T. (2008) 'SAP who? Inside one of SAP's smallest ERP customer success stories'. *CIO*, 25 September.

[94] Langley, N. (2003) 'No more "jam tomorrow"'. *Computer Weekly*, 3 November.

[95] Nestlé management report, 2008.

[96] Worthen, B. (2002) 'Nestlé's enterprise resource planning (ERP) odyssey'. *CIO*, 15 May.

[97] Nestlé investor seminar, 8 June 2005, chaired by Chris Johnson, responsible for the GLOBE programme.

[98] Berinato, S. (2003) 'A day in the life of Celanese's big ERP rollup'. *CIO*, 15 January.

[99] Celanese Corporation annual report, 2004.

[100] Flinders, K. (2008) 'Standard Life removes inflexible legacy systems to comply with FSA regulation'. *Computer Weekly*, 25 September.

[101] Wailgum, T. (2009) 'Now departing: Union Pacific's 40-year-old mainframe'. *CIO*, 12 August.

[102] Robb, D. (2004) 'Thrifty storage strategies'. *Computerworld*, 18 October.

[103] Hubbard, D.W. (2007) *How to Measure Anything: Finding the Value of 'Intangibles' in Business*. Hoboken, NJ: Wiley.

[104] Ibid, p. 10.

[105] Ibid.

[106] Savvas, A. (2008) 'Severn Trent uses interactive voice system to collect customer debts'. *Computer Weekly*, 12 September.

[107] Rossi, S. (2007) 'Qantas slashes design time using 3-D software from SolidWorks'. *Computerworld Australia*, 20 June.

[108] Lucas H.C. (1999) *Information Technology and the Productivity Paradox: Assessing the Value of Investing in IT*. New York: Oxford University Press, p. 12.

[109] Kaplan, R.S. (1986) 'Must CIM be justified by faith alone?' *Harvard Business Review*, March-April.

[110] Kaplan was quoting from Shewchuk, J. (1984) 'Justifying flexible automation'. *American Machinist*, October, p. 93.

[111] Ko, C. (2009) 'Check-in on Cathay Pacific with mobile phones'. *MIS-Asia*, 22 January 2009.

[112] Hubbard, *How to Measure Anything*, p. 23.

[113] Ibid, p. 86.

[114] Information obtained from Daimler Financial Services Americas website, at *www.daimler-financialservices.com/dcfs-na/0-876-571026-1-571307-1-0-0-0-0-0-10811-569676-0-0-0-0-0-0-0.html*, accessed on 24 December 2009.

[115] Overby, S. (2005) 'Lessons from Hurricane Katrina: it pays to have a disaster recovery plan in place'. *CIO*, 16 September.

[116] Lucas, *Information Technology and the Productivity Paradox*, p. 169.

[117] Black, F. and Scholes, M. (1973) 'The pricing of options and corporate liabilities'. *Journal of Political Economy*, Vol. 81, No. 3, pp. 637–54.

[118] Taudes, A., Feurstein, M. and Mild, A. (2000) 'Options analysis of software platform decisions: a case study'. *MIS Quarterly*, Vol. 24, No. 2, pp. 227–43.

[119] For example, see Lucas, *Information Technology and the Productivity Paradox*, p. 170.

[120] Similar points are made in Hubbard, *How to Measure Anything*, p. 266.

[121] Carr, 'IT Doesn't Matter', p. 6.

[122] 'The power of mobile money', *The Economist*, 25 September–2 October, p. 14.

[123] Brynjolfsson and Saunders, *Wired for Innovation*.

[124] Ibid, p. 43.

[125] Ibid, p. 44.

[126] Ibid, p. 58.

[127] Lewis, W.W. (2004) *The Power of Productivity: Wealth, Poverty and the Threat to Global Stability*. Chicago, IL: University of Chicago Press.

[128] Smith, A. (1776/2008) *The Wealth of Nations*. Radford, VA: Wilder, p. 9.

[129] Schumpeter, J.A. (1942) *Capitalism, Socialism and Democracy*. New York: Harper and Row, p. 83.

Index

A

B

C

F

Family Dollar Stores 80

Farwest Steel 62-3

Fleet Boston 106

flexibility 62-4, 89-91, 115, 128-9, 155-6

Ford, Henry 152

FoxMeyer Drug 119

framework

    applying 83-135

    important element in new 81, 135

    in practice 85-135

    introducing complexity 46-9

    need for a new one 3-23

    no standard 7

    towards a new one 31-82

future development 62-4 (*see also:* development)

G

Game Changer (book) 52

General Electric (GE) 36, 45-8

GLOBE (*see:* Nestlé SA)

H

Harris Corporation 50-1

Hershey 119

Honda 74-5

How to Measure Anything (book) 149

HSBC 115, 169

Hurricane Katrina 154

I

IBM 128

improving business 89-98, 150-3

Information Technology and the Productivity Paradox (book) 151, 156

innovation 13, 31, 52-3, 58, 69, 82, 96-8, 152, 167, 171-173

insights and implications 165-9

internet 18, 83, 131

integrating

    applications 106-14

    infrastructure 114-6

Intel 12-13

investors 73

'IT Doesn't Matter' (paper) 36, 41

J

jargon (*see:* abbreviations)

Java 18

JetBlue 63

John Deere 92

K

Kaiser Permanente 90

Kimberley-Clark 40

Kmart Corporation 119